T0224348

Computational Genomic Signatures

Synthesis Lectures on Biomedical Engineering

Editor
John D. Enderle, *University of Connecticut*

Lectures in Biomedical Engineering will be comprised of 75- to 150-page publications on advanced and state-of-the-art topics that spans the field of biomedical engineering, from the atom and molecule to large diagnostic equipment. Each lecture covers, for that topic, the fundamental principles in a unified manner, develops underlying concepts needed for sequential material, and progresses to more advanced topics. Computer software and multimedia, when appropriate and available, is included for simulation, computation, visualization and design. The authors selected to write the lectures are leading experts on the subject who have extensive background in theory, application and design.

The series is designed to meet the demands of the 21st century technology and the rapid advancements in the all-encompassing field of biomedical engineering that includes biochemical, biomaterials, biomechanics, bioinstrumentation, physiological modeling, biosignal processing, bioinformatics, biocomplexity, medical and molecular imaging, rehabilitation engineering, biomimetic nano-electrokinetics, biosensors, biotechnology, clinical engineering, biomedical devices, drug discovery and delivery systems, tissue engineering, proteomics, functional genomics, molecular and cellular engineering.

Computational Genomic Signatures
Ozkan Ufuk Nalbantoglu and Khalid Sayood
2011

Digital Image Processing for Ophthalmology: Detection of the Optic Nerve Head
Xiaolu Zhu, Rangaraj M. Rangayyan, and Anna L. Ells
2011

Modeling and Analysis of Shape with Applications in Computer-Aided Diagnosis of Breast Cancer
Denise Guliato and Rangaraj M. Rangayyan
2011

Computational Genomic Signatures
Ozkan Ufuk Nalbantoglu and Khalid Sayood

ISBN: 978-3-031-00522-0   paperback
ISBN: 978-3-031-01650-9   ebook

DOI 10.1007/978-3-031-01650-9

A Publication in the Springer series
SYNTHESIS LECTURES ON BIOMEDICAL ENGINEERING

Lecture #41
Series Editor: John D. Enderle, *University of Connecticut*
Series ISSN
Synthesis Lectures on Biomedical Engineering
Print 1930-0328   Electronic 1930-0336

Computational Genomic Signatures

Ozkan Ufuk Nalbantoglu and Khalid Sayood
University of Nebraska

SYNTHESIS LECTURES ON BIOMEDICAL ENGINEERING #41

ABSTRACT

Recent advances in development of sequencing technology has resulted in a deluge of genomic data. In order to make sense of this data, there is an urgent need for algorithms for data processing and quantitative reasoning. An emerging *in silico* approach, called computational genomic signatures, addresses this need by representing global species-specific features of genomes using simple mathematical models.

This text introduces the general concept of computational genomic signatures, and it reviews some of the DNA sequence models which can be used as computational genomic signatures. The text takes the position that a practical computational genomic signature consists of both a model and a measure for computing the distance or similarity between models. Therefore, a discussion of sequence similarity/distance measurement in the context of computational genomic signatures is presented. The remainder of the text covers various applications of computational genomic signatures in the areas of metagenomics, phylogenetics and the detection of horizontal gene transfer.

KEYWORDS

genome, Markov models, minimum description length, Kolmogorov complexity, phylogeny, classification, horizontal gene transfer, metagenomics, bioinformatics

Contents

Acknowledgments

This work was supported, in part, by a grant from the National Institutes of Health under grant K25AI068151. We thank Patricia Worster for her careful reading and critique of the manuscript.

Ozkan Ufuk Nalbantoglu and Khalid Sayood
May 2011

CHAPTER 1

Genome Signatures, Definition and Background

Since the discovery that deoxyribonucleic acid (DNA) is the primary repository of genetic information, understanding the molecular evolution of biological sequences such as DNA, ribonucleic acid (RNA) and proteins has been invaluable for understanding the driving forces, trends and implications of the evolution of species. Development of statistical tools for analyzing biological sequences has been useful for capturing the effect of evolution on genomes. An important discovery is that the compositional features of a genome carry information about the evolutionary history of a species.

These compositional features carry specific signals which permit organisms to be distinguished on the basis of genus and species. This specificity can be interpreted to be the result of the process of adaptation of the species to the environment. Environmental and structural parameters are some of the factors shaping DNA, RNA and protein compositions. Furthermore, physicochemical structural constraints and high-level cellular machinery also shape the organization of biological sequences.

Along with providing a means for distinguishing between species, the species specificity of these features can be used in a number of ways. The relative homogeneity of the compositional factors means that the species specificity of these features exists throughout the genome. These two properties of species specificity and pervasiveness are major components of a genomic signature.

1.1 DEFINITION OF COMPUTATIONAL GENOMIC SIGNATURES

Characterizations of species-specific features in biological sequences are often described by the term *signature*. The term genomic signature has been used homonymously corresponding to similar concepts but to different properties. For instance, a species-specific feature obtained from a genome is frequently used as a genome signature. Such a feature may be a short fragment of the genome unique to the organism. A sequence of around 20-25 bp in length has a low probability of appearing in all genomes. Therefore, those sequences are comprehensively searched for and labeled as barcodes belonging to specific taxonomic groups. Detection technologies, such as microarray platforms [Cannon et al., 2006] or polymerase chain reaction (PCR) assays [Livak et al., 1995, Phillippy et al., 2007], can detect these barcodes resulting in the detection of the unknown organism. This barcoding methodology has been used for building catalogues of species and identification of birds [Hebert et al., 2004], fish [Ward et al., 2005] and amphibians [Vences et al., 2005a] as well as a large set of other eukaryotes [Miller, 2007]. Similarly, barcoding using composition vectors gathered from rRNA sequences has also been used for similar purposes [Chu, 2006, Chu and Li, 2009]. The genomic signatures in this sense are located in a specific region of the genomes; and the knowledge of the entire genome, or at least the location and sequence of a specific region, is required for defining the genome signatures.

Unlike previous definitions of genome signature, computational genomic signatures utilize the relative homogeneity of genomes as well as the species specificity of DNA. A *computational genomic signature* is a species-specific mathematical structure that can be generated from an arbitrary genome fragment. Given a random fragment of sufficient length of any genome, one can generate the same (or similar) mathematical characterization as that obtained from the entire genome. The resulting structure is distinguishable from that obtained from the genome of a different organism. In order to introduce the distinguishability of signatures, a metric is also needed in the space where the signature is defined. That is:

$$d_S(S(G_{X_i}), S(G_{X_j})) < d_S(S(G_{X_i}), S(G_{Y_k})), \tag{1.1}$$

where G_{X_i} and G_{X_j} are random DNA sequences from the genome G_X, G_{Y_k} is a random DNA sequence from genome G_Y and $i, j, k \in \mathbb{N}^+$. $S(.)$ is an operation over the domain of possible DNA sequences and the range in a metric signature space. The distances in this signature space are shown with the metric $d_S(.,.)$. Ideally, the signature is embedded in any subsequence of a genome, that is $d_S(S(G_{X_i}), S(G_{X_j})) = 0$. In practice, due to the heterogeneities introduced by functional constraints and random mutations/deletions/insertions, these intergenomic distances are generally nonzero. These intergenomic distances depend on both the feature extraction ability of the signature and the metric defined in the signature space. Two attributes determine the quality of a genome signature: *species specificity* and *pervasiveness*. Genome signatures are pervasive in that they appear throughout the genome, and species specific in that they are different for different organisms.

1.2   COMPOSITIONAL FEATURES AS GENOME SIGNATURES

1.2.1   GC CONTENT:

GC content, a compositional feature of genomes discovered early, is a popular characterization which satisfies the genome signature definition. It measures the ratio of cytosine + guanine bases in a DNA sequence. The ratio of genomic GC content is biased across the tree of life and ranges from 16.5% (Carsonella ruddii) to 75% (Anaeromyxobacter dehalogens) [Nakabachi et al., 2006, Lee et al., 1956]. GC variation is also correlated with phylogenetic variation [Ochman and Lawrence, 1996].

The variation of GC content has been attributed to several factors. The difference in physico-chemical character of the cytosine–guanine and adenine–thymine bonds results in varying reactions to different factors. Examples include the fact that cytosine and guanine form three hydrogen bonds between the strands in the double helix as opposed to two hydrogen bonds for adenine and thymine which makes guanine-cytosine (GC) bonds more resistant to denaturation [Saenger, 1984], different reaction to reactive oxygen species damage [Marnett, 2000], the availability and lower cost of adenine/thymine (A/T) products, the preference of GC over AT in different respiratory behavior, growth temperatures and ecological conditions. Along with the selective perspective which maintains that GC bias is driven by selective pressures exerted by the environment, there is a neutralist camp which claims that [Freese, 1962, Sueoka, 1962] the bias is not a result of selection but is due to a neutral mutational behavior. Because of their variation with varying environmental parameters, GC content values appear to be species specific. Moreover, as the bases are distributed throughout a genome in similar proportions, the GC content satisfies the pervasiveness attribute of a genome signature.

The different values of GC content for various species and the relative conservation of GC content within a genome was noticed in the early 1960s [Sueoka, 1962]. We can observe GC content in randomly chosen 50 kbp genomic fragments of *Neisseria meningitidis* and *Mesorhizobium loti*, as shown in Figure 1.1 The GC content is also the simplest form of signature, since it consists of a single rational number and the distance metric is simply the arithmetic difference of these numbers.

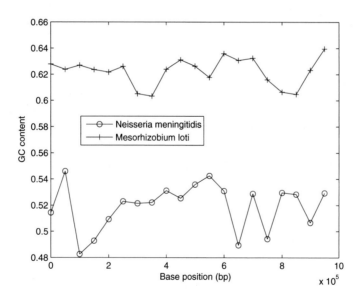

Figure 1.1: GC content of randomly chosen 50 kb genomic fragments of *Neisseria meningitidis* and *Mesorhizobium loti*.

1.2.2 AMINO ACID CONTENT:

The amino acid content is represented by a twenty dimensional vector of the relative frequencies of amino acids used in a protein or a proteome. It involves the simplest feature at the proteome level, analogous to GC content at the genome level. Certain organisms prefer different amino acids in their proteins, resulting in a spectrum of typical amino acid usage of various taxa.

It has been suggested that the species specificity of amino acid usage is the outcome of certain evolutionary processes. Response to different environmental temperatures [Cavicchioli, 2006, Hickey and Singer, 2004], economy of nutrient supply [Bragg and Hyder, 2004, Bragg et al., 2006], susceptibility to oxidation and the resulting behavior under different respiratory regimes [Berlett and Stadtman, 1997] are among the factors proposed for shaping the amino acid content.

The preference for certain amino acids is also fairly conserved throughout a genome. Because genes do not diverge significantly in the preference of amino acids they code for, this preference is pervasive through the genome. This signature property was used by Sandberg et al. [2003] for classification of proteins based on their amino acid content.

Table 1.1: F-scores of one-way-ANOVA for amino acid usage and synonymous codon usage. Distribution of profiles at different clade levels are considered.

F-value	synonymous codon usage	Amino acid usage
Genus	969.55	708.65
Family	1016.85	818.15
Order	1186.3	875.54

1.2.3   SYNONYMOUS CODON USAGE:

Synonymous codon usage is generally represented by 64 dimensional vectors which reflect the relative frequency of each codon coding for an amino acid. In the early 1980s, it was noted that each species systematically prefers certain codons to code an amino acid; this phenomenon is true for most genes of an organism [Grantham, 1980, Osawa et al., 1992, Gouy and Gautier, 1982, Xie et al., 1998]. The proposition that synonymous codon usage is species specific is known as Grantham's genome hypothesis.

The variation of synonymous codon usage among the genes of an organism is frequently attributed to gene expression levels and the relative abundance of tRNAs in a cell [Gouy and Gautier, 1982, Carbone et al., 2003]. Variation between genomes is more significant than intergenomic variation. Even though the usage of synonymous codons does not change the protein composition, it has also been linked to amino acid composition [D'Onofrio et al., 1991, Lobry, 1997, Foster et al., 1997, Sueoka, 1961], protein structure [Xie et al., 1998, Adzhubei, 1996, Gupta et al., 2000], directional mutational biases [Chen et al., 2004, Knight et al., 2001, Palidwor et al., 2010] and mRNA secondary structure [Zama, 1990]. The direct relationship of synonymous codon usage to environmental factors can be seen by the fact that synonymous codon usage carries signals revealing information about the thermal and respiratory behavior of an organism [Carbone, 2005].

Following a similar statistical methodology used for amino acid usage, it has been shown that synonymous codon usage exhibits genome signature characteristics [Sandberg et al., 2003]. Table 1.1 shows one-way ANOVA test results based on F-scores for amino acid usage and synonymous codon usage. Each gene is represented by its amino acid/synonymous codon usage profile, and analysis of variance is employed assuming each taxon as one group. The test was performed for the clade levels of genus, family and order. Higher F-scores imply a clearer separation of taxa in the vector spaces of the corresponding genome signatures.

1.3   METHODS OF CHARACTERIZATION EMBEDDED IN THE INITIAL WORK ON DNA

In the 1960s, the first mentions of genome signatures appeared as supplementary observations to experiments designed for different purposes. Before the birth of computational biology, and the existence of molecular databases and *in silico* genome analysis, [Kornberg et al., 1961, 1962] conducted a series of studies using the replication factors from phage ΦX 174 and primer sets to synthesize DNA from viral, bacterial, plant and animal sources. The ingenious technique they used involved $5' - P^{32}$ labeled DNA to obtain the percentage of different dinucleotides. Their main motivation, and thus the main observation, was confirming Watson-Crick base pairing by comparing the reverse complement doublets in forward and reverse strands. Along with achieving their primary goal, they also found that the frequency of occurrence of dinucleotides did not follow a random model. That is, the frequency of occurrence of a dinucleotide pair XY was not equal to the product of the frequency of occurrence of each individual nucleotide X and Y. They also found that the dinucleotide frequencies obtained from different taxonomies, such as mouse tumors, crab testis, bovine liver, plants and viral DNA, were distinguishable by dinucleotide frequencies. In particular, they found that the frequency of occurrence of the CpG dinucleotide fits a random model for bacteria; but it moves progressively away from a random model for echinoderms and vertebrates. Another important observation they reported was that the synthesized DNA sequences had the same doublet frequency characteristics with the primers used to synthesize these sequences for viral, bacterial and animal sources. These additional observations are actually indications of the species specificity and pervasiveness of doublet frequencies as genome signatures. [Subak-Sharpe, 1967a, Subak-Sharpe et al., 1977] defined the term *general design of an organism* as the normalized frequency of occurrence (odds ratio) of dinucleotides, and they noted the similarity of the general design of several small mammalian viruses and their hosts [Subak-Sharpe et al., 1967b].

1.4   DINUCLEOTIDE ODDS RATIO AS A GENOME SIGNATURE

After the first indications of the existence of genomic signatures, it took almost 30 years to reconsider the concept. With the increasing availability of genomic sequences, Karlin and colleagues, in a sequence of papers [Karlin et al., 1992, 1994a,b,c,d, 1995, 1997a,b, 1998], extended the work of

Kornberg et al. and Subak-Sharpe et al. and coined the term *genomic signature*. Initially, the odds ratio of dinucleotides (along with tri- and tetranucleotides) to measure the divergence of neighboring bases from expected distributions was introduced to observe the over- and underrepresentation of dinucleotides in genomes [Karlin et al., 1992]:

$$\rho_{XY}^* = \frac{f^*(XY)}{f^*(X)f^*(Y)}. \tag{1.2}$$

Here, $f^*(XY)$ stands for the frequency of the dinucleotide XY in the given fragment concatenated with its reverse strand. The frequencies of the bases X and Y are $f^*(X)$ and $f^*(Y)$. This odds ratio gives an overrepresentation or an underrepresentation measure for all 16 dinucleotides. Note that $f^*(X)$ values are calculated using both strands. Because of the Watson-Crick pairing, $f^*(G) = f^*(C) = f(G + C)$, and the same property applies for A and T. The frequencies without star superscripts are the frequencies calculated using one strand of the genome. Initially, these measurements were used individually; and global properties of dinucleotide occurrence, such as the underrepresentation of AT in almost all taxonomies, underrepresentation of CG in vertebrates and mitochondrial DNA, and overrepresentation of homodimers, along with the corresponding evolutionary implications, were discussed. [Karlin et al., 1994b] examined the normalized frequency of occurrence of di-, tri- and tetra-nucleotides in various eukaryotic genomes. As a result of this study, they noted that the Euclidean distance of relative abundance profiles for closely related organisms were smaller than the distances calculated for phylogenetically distant organisms. Later on, a metric which took into account all 16 dinucleotide abundance values, the δ distance, was introduced [Karlin et al., 1995]:

$$\delta^*(f, g) = 1/16 \sum_{XY} |\rho_{XY}^*(f) - \rho_{XY}^*(g)|. \tag{1.3}$$

Having provided the two requirements for a genome signature, the signature and the distance metric in signature space, Karlin et al. investigated the species specificity and pervasiveness of that signature. It was seen that the δ distance is very small within the same species, being only 2-3 times that of the distances between random sequences. Another result was that within the genome, the distance is generally smaller than the intergenomic measurements. In fact, in some cases, the species specificity and pervasiveness of dinucleotide relative abundance ratio profiles are even visible to the naked eye without any metric definition. An example is shown in Figure 1.2 for 20 random 50 kbp segments from *Neisseria meningitidis* and *aquifex aeolicus* genomes. Clearly, the 50 kbp sections are distinguishable for these two genomes.

A fruitful series of applications followed this initial discovery of genome signatures. The dinucleotide abundance signature along with the δ distance was observed to be pervasive in eukaryotes for genomic fragments of length greater than 50 kbp. Moreover, according to their genome signature analysis, archea appeared to be an inconsistent clade having large signature distance between the members. Although the dinucleotide abundance profiles of nuclear DNA and mitochondrial DNA are significantly different from each other, it was found that the distances between the mitochondrial

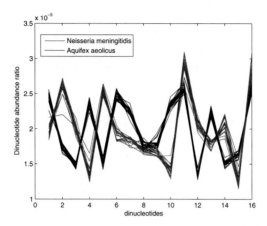

Figure 1.2: The dinucleotide odds ratio profiles for 20 random 50 kbp segments from *Neisseria mening-itidis* and *aquifex aeolicus* genomes.

DNA are similar to the distances obtained from nuclear DNA segments. This result was considered quantitative evidence for the coevolution of eukaryote cells and their mitochondria. Moreover, the mitochondria of mammals were reported as being very similar to each other, while animal and fungal mitochondria DNA were moderately similar and all were very different from plant and protist mitochondrial sequences. With their genome signature studies on virus and bacterial plasmids, Karlin and colleagues found that both virus and plasmids resemble the structure of their hosts. Also among viral genomes, single-stranded RNA viruses were found to be the species having the most obscure signatures which are close to random sequences. They attributed that random nature to the high mutation rate of single-stranded RNA.

During their investigation of genomic signatures, Karlin and colleagues were not able to determine a clear relationship between environmental factors (e.g., habitat propensities, osmolarity tolerance, chemical conditions) and genome signature. For the most part, they attributed the emergence of signatures to the structural properties of the DNA polymer, such as dinucleotide stacking energies, curvature, chromosomal organization, DNA packaging, DNA replication, transcription and repair mechanisms.

1.5   CHAOS GAME REPRESENTATION

The history of genome signature discovery has developed along two different lines from two different biological sequence analysis groups. The first group contains the initial *in vivo* approaches investigating dinucleotide occurrence frequencies. In this approach, the over- and underabundance of nucleotide doublets accounts for species specificity and pervasiveness. Another branch of sequence analysis followed statistical mechanics approaches to analyze the genomic sequences, finally ending up with another form of genome signature. Later on, the tight connection between those two concepts in that they were both functions of oligonucleotide composition was reported.

The attempts to represent genome sequences in other mathematical forms, in which a rich repertoire of analysis tools is available, has been of great interest to researchers. Some of these approaches have their roots in statistical mechanics. Representing the sequences as random walks [Bai et al., 2007, Gates, 1985, Leong and Morgenthalar, 1995, Kowalczuk et al., 2001] has revealed some features, such as the walks of DNA sequences resemble fractal behavior. Moreover, divergence from random sequences and exhibiting Markov-like behavior provided a basis for further investigation of compositional features. In 1990, [Jeffrey, 1990] proposed a method he called the Chaos Game Representation (CGR) to visualize the genomic sequences. This was a method employed from nonlinear dynamics [Devaney, 1989] as a two-dimensional representation of symbolic sequences. According to this scheme, a symbolic sequence is scanned with a running window of length k; and with every step the observed k-mer is represented in a two-dimensional iterated map. Simply, we can assume that from the left-top quadrant in clockwise direction each quadrant represents C, G, T, and A, respectively, in a square. The first base is placed in the corresponding quadrant, after that the quadrant is divided into four quadrants and the same procedure is applied for the second base. Iteratively, the observed window finds its place in one of the 4^k squares, in k steps of iteration. Complex nonrandom symbolic sequences are observed to form fractal images with chaos game representation. Jeffrey observed this behavior in DNA sequences and concluded that DNA sequences were far from random.

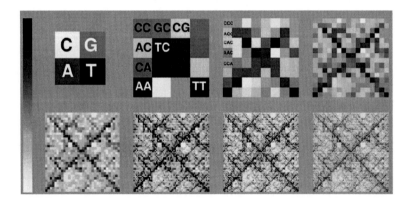

Figure 1.3: The generation of the CGR of *Archeoglobus Fulgidus* genome in 8 iterations (figure taken from [Deschevanne et al., 1999]).

An objection to chaos game representation of genomic sequences arose from Goldman, claiming that it reflects the short term correlations of DNA rather than capturing complex structures. He added the claim that the same images can be generated from mono-, di- and trinucleotide frequencies of DNA sequences. Indeed, he was able capture the "double scoop" character, an indication of scarcity in a CG doublet, of CGR observed in vertebrate genomes and in vertebrate viruses. Goldman was right in claiming that CGR images do not capture complex structures but reflect the short-term correlations, and he was wrong in claiming that those images do not provide superior information to that obtained from oligonucleotide frequencies up to trinucleotide or even codon usage. In fact, CGR images contain the information of k-mers and not more than that. Since the correlations in DNA are longer than dependencies three base apart, CGR can provide better knowledge than codon usage. To see how CGR images contain exact k-mer frequency information, we can follow this reading: the idea of this representation is that the whole set of frequencies, from mononucleotide to k-length word, found in a given genomic sequence can be displayed in the form of a single image in which each pixel is associated with a specific word. The difference of this specific representation from a random arrangement of pixels in the image is that the generation of the image is a recursive process starting from 4 pixels for mononucleotides and splitting each pixel by 4 in every iteration for each word length expansion. This can be thought of as increasing the resolution of a quantized image. The grayscale value indicates the relative frequency of a word. In Figure 1.3 the generation of the CGR of *Archeoglobus Fulgidus* genome is shown. The resulting images show certain characteristics as the word length increases. A human expert can comprehend the characteristics of a genome by analyzing the CGR. For instance, the lighter upper part indicates low G+C composition and diagonally oriented lines indicate the abundance of purine and pyrimidine stretches. These diagonal lines can be seen in Figure 1.3

It was [Deschevanne et al., 1999], who discovered that CGR representation could also be used as a signature. With CGR images created from different organisms, it was clear that different organisms attain distinguishable CGR images. Moreover, the images obtained from random genomic fragments, down to 1000 bp in length formed images resembling different variations of the same image to the human eye (Figure 1.4).

Heuristically, the pervasiveness and species specificity of CGR images are visible. However, as signatures are mathematical structures, there is a need for metrics to quantify the signature behavior as mentioned before. Euclidian distances between the CGR images, obtained by adding the squared pixel differences for the same pixel locations, were calculated; and the species specificity and pervasiveness were shown by computational experiments [Deschevanne et al., 2000]. It is clear that the Euclidian distances of 2^k times 2^k CGR images correspond to the vector distances of k-mer frequencies in the composition space. Therefore, oligonucleotide frequencies are genome signatures. A close relationship between the dinucleotide abundance ratio signatures and GCR images was noticed by [Wang et al., 2005]; and it was concluded that as the information of dinucleotide abundance profiles is already embedded in CGR images, they belong to a spectrum of genomic signatures.

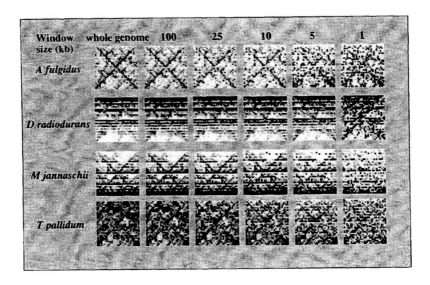

Figure 1.4: CGR images for A fulgidus, D radiodurans, M jannaschii and T pallidum for varying fragment length (figure taken from [Deschevanne et al., 1999]).

These results imply that, dinucleotide abundance ratios, CGR and oligonucleotide frequencies are computational genomic signatures of the same class.

1.6   A UNIFIED FRAMEWORK OF GENOME SIGNATURES: FUNCTIONS OF OLIGONUCLEOTIDE OCCURRENCE

It is possible to define a general compositional feature from which the genome signatures defined above can be deduced, namely, the frequency of oligonucleotide occurrence in a DNA fragment. GC content, synonymous codon usage and amino acid content can be approximately expressed as functions of oligonucleotide frequency profiles. Moreover, genome signatures defined on dinucleotide abundance ratios and CGR images are functions of oligonucleotide frequencies.

Given an oligonucleotide frequency vector of a DNA sequence, the GC content of this sequence can be obtained by the summation:

$$f(GC) = \sum_{GC} \sum_{x_1} \sum_{x_2} \cdots \sum_{x_{k-1}} f(x_1 x_2 .. x_k) \tag{1.4}$$

This is basically a linear projection on a line in the oligonucleotide frequency space which can be represented as the dot product of a vector with 4^k entries of 0's and 1's with an oligomer frequency vector. We can represent this mapping with P_{GC} and the mapping operation as $f(GC) = P_{GC}(f(X))$, where $f(GC)$ is the GC content, X is the k-mer relative frequency vector and $P_{GC}(.)$ is the linear function.

Summing up a trimer DNA composition vector with the help of the standard genetic code, we can obtain the approximate amino acid content vectors with a linear projection represented by a 20 × 64 binary matrix. Although the relative frequency of an amino acid equals the codon frequencies coding it, we substitute the codon frequencies with trinucleotide frequencies in order to obtain the relationship. The codon frequencies are calculated with a moving window of three bases, while the trinucleotide frequencies do not take the reading frames into account and average the frequencies over all reading frames. That is why this is an approximate mapping. The representation of this mapping is $(P_{aa} \circ P_{k3})$; and the mapping operation is $f(aa) \approx P_{aa}(P_{k3}(f(X))$, where $f(aa)$ is the amino acid content, $f(X)$ is the k-mer relative frequency vector, P_{aa} is the linear function mapping trimers to amino acid frequencies and P_{k3} is the linear function mapping k-mer frequencies to trimer frequencies. The error resulting from the approximation is negligible ($r^2 = 0.9987$, $P < 0.0001$).

Clearly, synonymous codon usage can be obtained by normalizing absolute codon frequencies (which are approximately the trinucleotide vectors) with the amino acid content. Both are linear projections in the oligonucleotide content space, which results in a nonlinear mapping within this space. The representation of this mapping is $(P_{scu} \circ P_{k3})$; and the mapping operation is $f(X_{scu}) \approx P_scu(P_{k3}(f(X)))$, where X_{scu} is the vector containing synonymous codon usage, X is the k-mer relative frequency vector, P_{scu} is the nonlinear function mapping trimer frequencies to synonymous codon usage vectors and P_{k3} is the linear function mapping k-mer frequencies to trimer frequencies. There is a strong correlation between the approximate mapping and the actual synonymous codon usage values ($r^2 = 0.98$, $P < 0.0001$).

1.7   NATURAL SELECTION OF PROTEINS VS. DIRECTIONAL MUTATIONAL PRESSURES

The fundamental categorization of genomic sequences is in terms of coding and noncoding DNA. The coding DNA contains the genes which are directly linked to structural and metabolic elements through protein synthesis. Noncoding DNA, on the other hand, does not directly take a role in transcription and, therefore, does not have a one-to-one mapping with proteins of the organism. Based on this separation, it is proposed that the compositional structure of DNA sequences can be driven by two major kinds of selective processes. First is the direct mutational pressure of intrinsic and environmental factors on genomes which modify the sequences. The second process is based on the natural selection of proteins and RNA sequences which imposes a selective pressure on certain protein and messenger RNA compositions and, thus, indirectly shape the RNA and DNA sequences (Figure 1.5).

The driving force behind molecular evolution has long been a mystery. The most plausible explanation could be a combination of both selective forces acting in cooperative or competing terms, since it is not possible to reject any of the selective processes with current observation. However, a number of recent studies form a basis for supporting the hypothesis that selective forces acting on DNA form the dominant component of molecular evolution. In this section, we discuss the evidence supporting this hypothesis.

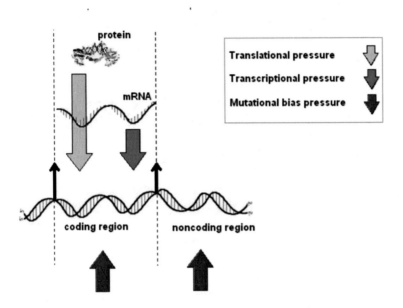

Figure 1.5: The selective pressures on coding and noncoding regions of a genome. Selective pressures associated with protein synthesis (e.g., translational, translational pressures) affect on coding regions. On the other hand, directional mutational pressures are global throughout the genome.

An interesting study investigating codon usage biases [Chen et al., 2004] reported that synonymous codon preference can be predicted from intergenic composition, which implies that codon usage is driven mostly by genome-wide mutational pressures. If both amino acid content and codon usage are driven by GC content, and, consequently, the mutational biases, an indirect correspondence of the synonymous codon usage and amino acid usage might, emerge. Another principal observation in that study is that when the organisms are represented in the codon space, the first two principal components of the data are associated with the GC content and the optimal growth temperature of the species, which explains most of the codon usage variance. The codon usage of a species is predicted from the GC content of the intergenic region and a Markov model is trained on the non-coding sequences, which can successfully reconstruct the first two principal components of codon usage vectors in the codon space.

Two other studies build models of codon usage that describe the codon usage as a function of genomic GC content [Knight et al., 2001, Palidwor et al., 2010]. Their basic methodology involves regressing context-dependent models between the GC content and the codon usage. Moreover, the amino acid composition is also defined as a function of the GC content. These results imply that the codon usage and amino acid composition are roughly determined by factors that affect the genome globally.

Similar arguments can be advanced for the other studies investigating amino acid [Tekaia and Yeramian, 2006] and synonymous codon preference [Carbone, 2005] in projected spaces. In [Tekaia and Yeramian, 2006], 20 dimensional feature vectors containing amino acid usage are projected onto a correspondence analysis plane which is associated with GC content (explains 63% of the data) and the optimal growth temperature (explains 14% of the data). Similarly, in [Carbone, 2005], 64 dimensional synonymous codon usage vectors are projected onto the first two principal components which are correlated with GC content (explains 45% of the total variance) and the optimal growth temperature (explains 13% of the total variance). Note that the translational selective pressure on the usage of synonymous codons is not effective whereas the usage of amino acids can be primarily restricted by translational pressures. However, the basic factors shaping the codon usage and amino acid compositions are reported to be similar. This could result in two interpretations: first, environmental and structural factors shaping the DNA with direct mutational pressures are in accordance with the natural selection of proteins and RNA which constrain the composition of amino acid sequences. Second, the factors shaping DNA directly with mutational pressures are dominant to protein selection and dictate the composition of proteins. In both cases, the effect of mutational pressures has greater visibility since they imply the recessivity of translational or transcriptional pressures.

CHAPTER 2

Other Computational Characterizations as Genome Signatures

To this point, we have concentrated on compositional properties of genomes in the form of short oligonucleotide frequencies; and we have shown that oligonucleotide frequencies constitute genome signatures. Following the genome analysis literature from different disciplines, two different branches of research came up with the concept of genome signatures. Being able to capture genome-specific structures without any need of entire chromosomes, but sufficiently long arbitrary genomic sequences, both dinucleotide abundance ratios and chaos game representations belong to the same class of signatures. They are both deducible from oligonucleotide frequency vectors and are instances of projections in the component space as discussed previously.

In this chapter we explore other computational structures which can be used as genome signatures. Armed with the definition of species specificity and pervasiveness, it is possible to systematically propose mathematical structures and test them for the presence of these two characteristics. In practice, the specificity and pervasiveness of signatures are not ideal [Jernigan and Baran, 2002] and as the genomic fragments get shorter, the signature derived from these fragments diverges from the actual signature of the genome. This deviation varies according to the strength of the pervasiveness of different signatures. Moreover, genomes belonging to evolutionarily close organisms might be indistinguishable by genome signatures and the distinguishability might vary according to the signature defined. Although, there is no benchmark using which we can definitively declare a mathematical structure to be a signature based on its specificity and pervasiveness, we can employ statistical tests on genome fragments of varying length, and by comparing the results of different signatures can obtain an idea of the strength of a signature. This is the methodology we follow to compare different signatures; and, since the mathematical representations we present here have competitive statistics, we refer to all of them as genomic signatures.

As in the case of the signatures described previously, some of the computational characterizations we discuss in this chapter are tightly connected to oligonucleotide frequencies; and thus they can also be considered as projections in the composition space, i.e., functions of genome composition statistics. The remaining characterizations cannot be directly connected to the oligonucleotide frequency of occurrence vector, although there is dependence on composition and preference/avoidance of certain words.

2.1   LONG-TERM CORRELATION STATISTICS AS GENOME SIGNATURES

Oligonucleotide frequency vectors consist of 4^k (k being the length of the oligomer) components, each component being the the frequency of occurrence of a specific k–mer. The exponential growth of the number of parameters as a function of k restricts the length of k-mers to small values. If the number of parameters is much larger than the size of the DNA fragment, as would be the case for typical DNA sequences, the estimation of frequencies of occurrence for larger values of k will lead to overfitting. Therefore, k has to be small and oligonucleotide vectors are capable of capturing only the short-term dependencies in genomes. The genome signatures, which are variants of oligonucleotide content (e.g., dinucleotide abundance ratios, chaos game representations), possess their signature characteristics due to dependencies between nearby nucleotides. Thus, characterizations of short-term correlations can be used to define genome signatures. An immediate question is *"Are long-term correlations also specific to a genome and homogenous?"* This implies that if we could measure long-term correlations in a genome, they might exhibit properties of computational genomic signatures. The answer to that question is not obvious, since it is not possible to find an intuitive rationale to propose conserved long-term correlations in genomes. Only an empirical attempt to measure genomic correlation will do.

To measure the correlation of a time series or a random process when the elements of the series are real valued is a well studied topic. With the assumption that the correlation and statistics of the sequence are conserved (i.e., wide sense stationary), the autocorrelation function can be estimated as:

$$\hat{r}(k) = \frac{1}{(n-k)\sigma^2} \sum_{t=1}^{n-k} (x_t - \mu)(x_{t+k} - \mu). \tag{2.1}$$

Here, x_t, n, μ, σ are the numeric sequence, its length, mean and variance, respectively. If biological sequences were real valued, the profiles of $\hat{r}(k)$ could be tested for their signature characteristics. However, estimating the correlations of symbolic sequences is not as straightforward. Estimating the correlation of symbolic sequences requires either a mapping of the symbolic sequence to a numerical sequence or the use of models to represent the genomic sequences as symbolic random processes. The former has been used for representing DNA sequences as random walks. For instance, for the DNA walk in which the walk is incremented by +1 if the next symbol is a pyrimidine base and decremented by 1 if it is a purine base, the mean square fluctuation observed is different from what one would expect from random sequences or Markov models [Peng et al., 1992]. This indicates that long-term correlations exist in DNA. The existence of this long term structure has since been validated in various studies [Li and Kaneko, 1992, Voss, 1992, Karlin, 1993, Buldyrev et al., 1995, Holste et al., 2003].

Investigating the correlations in genomes by mapping the DNA sequences into numerical data could give us an idea about the existence of such dependencies. However, the results are dependent on the mapping; and there is no trivial way of defining a DNA to number mapping. Instead, attempting stochastic sequence analysis in the native domain could mitigate most problems resulting from the representation of DNA as numeric sequences. We can do so using concepts from information theory. Several powerful sequence analysis schemes have been developed for searching for correlations in DNA sequences using various information theory tools [Herzel et al., 1994, Holste et al., 2000, Grosse et al., 2002, Berryman et al., 2004]. We will first introduce an approach proposed by Dehnert et al. to estimate the long term correlations of DNA to be utilized as genomic signatures.

2.1.1 DNA AS AN AUTOREGRESSIVE PROCESS

In a discrete autoregressive stochastic process, a symbol being emitted at time t is a function of the previous symbols. Therefore, it has memory; and by this memory short-, mid- or long-range correlations are introduced. In most systems, longer range correlations die out and become negligible in practical terms. Therefore, the memory, or the order of the systems, can be limited in practice. Autoregressive processes can be defined in terms of symbolic sequences. Such a model is called a discrete autoregressive process ($DAR(p)$) [Jacobs and Lewis, 1978, 1983]. In this case, a DNA sequence is assumed to be a $DAR(p)$ [Dehnert et al., 2003] as

$$x_n = V_n x_{n-A_n} + (1 - V_n) y_n. \tag{2.2}$$

Here, x_n is the nth symbol in a DNA sequence ($x_n \in \{A, C, G, T\}$). V_n is a Bernoulli process taking a value of one with probability ρ and zero with probability $1 - \rho$. A_n is an integer in $\{1, 2, 3, \ldots, p\}$, attaining each value with the probability $\alpha_1, \alpha_2, \alpha_3, \ldots, \alpha_p$. y_n is another random process with independent and identically distributed probabilities of attaining an element from $\{A, C, G, T\}$, represented by the marginal distribution π.

The process can be interpreted as follows. A new symbol in a DNA sequence is either picked from one of the previous p symbols before it, or emitted independently. Therefore, V_n works as a switch between random generation and selecting a symbol from near history. This solely depends on the probability ρ. If ρ is zero, there are no dependencies between the nucleotides and DNA is a random sequence and can be modeled as a zeroth-order Markov model. At the other extreme, each base depends on a context of length p. When the new symbol is picked from the previous p symbols, the probability α_i determines which one is selected. Note that α_i is the conditional probability of x_n being equal to x_{n-i} given x_n is selected from the history of the sequence. Therefore, it can be used as an indicator of the dependencies between bases i positions apart in the sequence. Therefore, the parameter vector $\alpha = [\alpha_1, \alpha_2, \alpha_3, \ldots, \alpha_p]$ can be used as the genome signature reflecting the dependencies of dinucleotides placed up to p bases apart.

Given the parameters of the $DAR(p)$ model, a simulated DNA sequence can be generated. However, to utilize this computational tool to define the corresponding genome signature, we have to estimate the parameters given a DNA sequence. Dehnert et al. uses a version of Yuke-Walker estimation to obtain the required parameters. According to this, the autocorrelation function of the $DAR(p)$ process can be represented with the Yuke-Walker equations:

$$r(k) = \rho\alpha_1 r(k-1) + \rho\alpha_2 r(k-2) + \ldots + \rho\alpha_p r(k-p), k \geq 1. \tag{2.3}$$

Expressing this as a system of linear equations:

$$r(1) = \rho\alpha_1 r(0) + \rho\alpha_2 r(1) + \ldots + \rho\alpha_p r(p-1)$$
$$r(2) = \rho\alpha_1 r(1) + \rho\alpha_2 r(0) + \ldots + \rho\alpha_p r(p-2)$$
$$\vdots$$
$$r(p) = \rho\alpha_1 r(p-1) + \rho\alpha_2 r(p-2) + \ldots + \rho\alpha_p r(0)$$

Given the autocorrelations, the linear system can be solved and $\alpha = [\alpha_1, \alpha_2, \alpha_3, \ldots, \alpha_p]$ can be obtained. It was shown that the ad hoc autocorrelation estimator performs well with symbolic sequences [Jacobs and Lewis, 1983]. In this case, the autocorrelation function is:

$$\hat{r}(k) = 1 - \sum_{a_i \in A} B_m(k, a_i) \frac{1}{1 - \pi(a_i)}. \tag{2.4}$$

Here $A = \{A, C, G, T\}$, and the function B_m is

$$B_m(k, a_i) = \frac{1}{m-k} \sum_{a_i \neq a_j \in A} \sum_{l=1}^{m-k} \delta_{a_i}(x_l)\delta_{a_j}(x_{l+k}) \tag{2.5}$$

where $\delta_a(x) = 1$ when $a = x$ and 0 otherwise.

With this version of the Yuke-Walker estimation of $DAR(p)$ model parameters, the estimated vector $\alpha = [\alpha_1, \alpha_2, \alpha_3, \ldots, \alpha_p]$ has been used as a genome signature and utilized for modeling eukaryote chromosomes and measuring the distances between the chromosomes of the same organism and the chromosomes from different organisms [Dehnert et al., 2005a, 2006]. In Figure 2.1, the plots of α vectors of up to 30 components are illustrated for all chromosomes of 6 eukaryotic organisms.

It is clear to the human eye that while the intergenomic parameter vectors are very similar, the pattern is different for different organisms. That implies species specificity and pervasiveness of α vectors or, namely, the correlation strength [Dehnert et al., 2005a]; thus, it is a genomic signature. This signature has been reported to be specific, but it becomes hard to distinguish between the signatures of close species. Using an ℓ_1 metric (i.e. $d(\alpha_1, \alpha_2) = \sum_i |\alpha_1(i) - \alpha_2(i)|$) to measure the distance of signatures, it was observed that the chromosomes of human and chimpanzees are difficult to distinguish from each other.

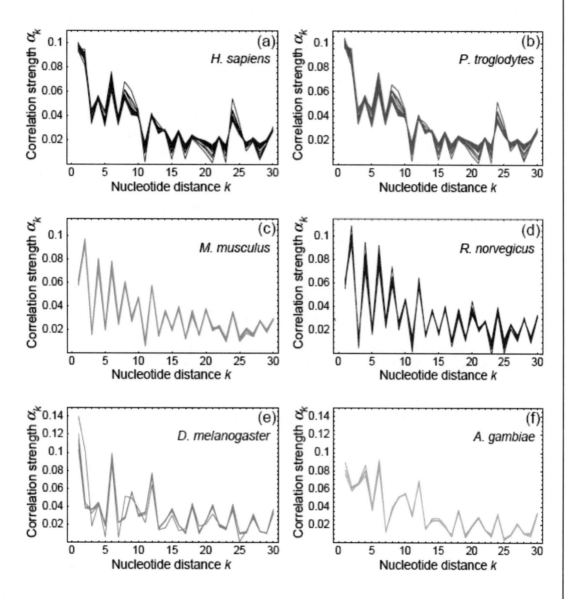

Figure 2.1: Correlation strength profiles for 6 different eukaryotic species (H. sapiens, P. troglodytes, M. musculus, R. norvegious, D. melanogaster, and A. gambiae). First 30 components are plotted, and the profiles of all chromosomes are shown.

2.1.2   AVERAGE MUTUAL INFORMATION PROFILES

Another method of detecting long-range correlations in DNA sequences is again derived from information theory. The average mutual information (AMI) measure was first introduced in the area of digital communications for the study of signals under noisy channel conditions [Shannon, 1948]. It has been used in computational biology for understanding dependent events like correlated mutations at noncontiguous sites [Korber et al., 1993] and secondary structures and correlations in protein sequences [Roman-Roldan et al., 1996, Giraud et al., 1998, Herzel and Grosse, 1997, Hofacker et al., 2002, Lindgreen et al., 2006].

Assume x is a random process emitting the DNA sequence, and the bases x_i and x_j at the positions i and j have marginal distributions $P(x_i)$ and $P(x_j)$. If there is correlation between bases x_i and x_j, the bases are not statistically independent. Therefore, the marginal distributions have shared information introduced by the base dependencies; and this shared information can be measured using the notion of mutual information. This measure gives us the information we would have about base x_i if we knew base x_j or vice versa. This can be represented as a reduction in the uncertainity about x_i given the random variable x_j. Measuring the uncertainty by Shannon entropy, the mutual information becomes:

$$I(x_i; x_j) = H(x_i) - H(x_i|x_j)$$
$$= H(x_i) - (H(x_i, x_j) - H(x_j)) \tag{2.6}$$

where $H(.)$ is the Shannon entropy. In the case of AMI, a DNA sequence is again viewed as a stochastic process with the assumption that the process is wide sense stationary and ergodic. This means that the information about the base distributions can be estimated from the DNA fragments, and the information is not position dependent. Therefore, the entropies can be estimated over x, and an average information can be assigned for a nucleotide pair placed k bases apart for all $|j - i| = k$. Thus:

$$I(x_i; x_j) = I(x; x(k)) = I(k) \tag{2.7}$$
$$= H(x) + H(x(k)) - H(x, x(k))$$
$$= -\sum_i P(x_i) \log_2(P(x_i)) - \sum_i P(x_{i+k}) \log_2(P(x_{i+k})) + \sum_i P(x_i, x_{i+k}) \log_2(P(x_i, x_{i+k}))$$
$$= \sum_i P(x_i, x_{i+k}) \log_2\left(\frac{P(x_i, x_{i+k})}{P(x_i)P(x_{i+k})}\right) \tag{2.8}$$

The probability estimates can be obtained using relative frequency counts of pairs of bases located k base pairs apart. The average mutual information gives a statistical measure of how much information is shared between nucleotides k bp apart; therefore, it forms a measure of the correlation within a DNA sequence. When the profile of a set of location distance values, $[I(1)I(2)\ldots I(n)]$ is compiled, it forms an average mutual information profile (AMI profile), providing the dependencies in short-, mid- or long-range. AMI profiles appear to be different for varying species and the signatures obtained from different parts of organisms resemble each other. [Bauer et al., 2008] investigated the signature behavior of AMI profiles and observed that much like correlation strength, AMI profiles are similar for different chromosomes of the same eukaryotic organism, while they show different patterns for each of those organisms. In Figure 2.2, AMI profiles to $n = 50$ are plotted for all the chromosomes of four eukaryotic organisms; the species specificity and pervasiveness can be observed graphically on that figure.

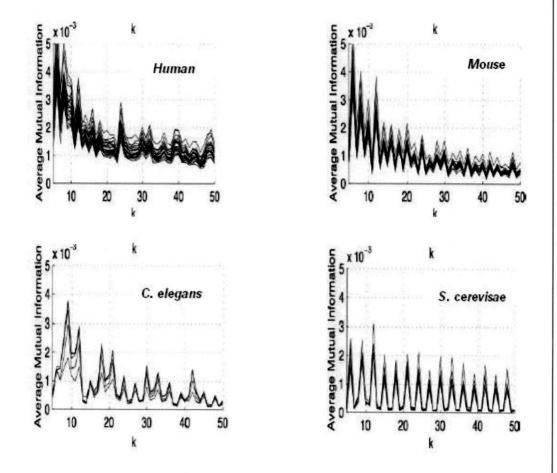

Figure 2.2: AMI profiles for the first 50 components are plotted for all the chromosomes of four eukaryotic organisms (mouse, human, C. elegans, S. cerevisae).

We have argued that most genomic signatures belong to the same class, because they are deducible from long oligonucleotide counts and are linear projections or nonlinear mappings in a high dimensional composition space. Even though, there is such a relationship between the correlation strength signature and the composition space, it is not explicit since the estimation of correlations

Table 2.1: F-scores of one-way-ANOVA for AMI profiles. Distribution of profiles for varying fragment length at different clade levels are considered.

F-value	1 kbp	10 kbp	50 kbp
Genus	105.75	708.65	2215.9
Family	120.08	818.15	2544.4
Order	127.28	875.54	2664.7

are performed via the parameter estimation of a discrete autoregressive model. On the other hand, this relation can still be claimed for AMI profiles, viewing $I(k)$ as the average log odds ratio of dinucleotide relative abundance, where the dinucleotides are located k bases apart from each other. We can show that the corresponding frequencies can be obtained by a linear mapping of the composition space. Assume the oligonucleotide frequency vector for n-mers:

$$f(x_1 x_2 \ldots x_n) = \begin{bmatrix} f(AAA \ldots A) \\ f(AAA \ldots C) \\ \vdots \\ f(TTT \ldots T) \end{bmatrix}. \tag{2.9}$$

the AMI profile is calculated over the dinucleotide frequencies. It is possible to deduce the dinucleotide frequencies from the genome signature by aggregating entries by summing them up. The resulting vector with 16 entries is:

$$f(x, x(k)) = \begin{bmatrix} f(xx \ldots xAx \ldots xAx \ldots x) \\ f(xx \ldots xAx \ldots xCx \ldots x) \\ \vdots \\ f(xx \ldots xTx \ldots xTx \ldots x) \end{bmatrix}. \tag{2.10}$$

Here, x denotes a "don't care condition" which represents any of the four bases. Clearly, $(k-1)$ $f(x, x(k))$ vectors can be generated from the oligonucleotide count vectors by summing up the "don't care" components x. This operation corresponds to a matrix multiplication of the n-mer frequency vector with a $n \times 4^n$ vector of 1s and 0s. Therefore, the AMI profile can be considered to be a nonlinear function of a linear projection in the component space which, consequently, is a nonlinear mapping of the component vectors. However, this theoretical result does not have an important implication in practice; because the required dimension for that mapping requires a very high number of parameters which cannot be estimated with realistic sized DNA fragments. For example, the number of parameters for an oligonucleotide vector to deduce an AMI profile of 30 variables is around 360 million fold greater than the total length of the human genome, which would result in overfitting with relative frequency counts. Therefore, both measures can be assumed to belong to a different class of genome signatures that utilize the longer range correlations in genomes.

Although it is possible to discuss the causes of relatively shorter dependencies in genomes, the factors resulting in the conservation of longer range correlations in DNA are not well comprehended. They are mostly attributed to the structural properties of DNA such as supercoiling and the corresponding 10-11 bp periodicities [Herzel et al., 1999, Trifonov and Sussman., 1980]. Also, Alu and SINE repeats [Berryman et al., 2004], and tandem repeats are thought to result in long-term correlations. However, removing all annotated repeats and investigating the correlation strength signature, it is still possible to observe the intragenomic similarities [Dehnert et al., 2005b], which might imply that structures other than well-known repeats are involved in the long-range correlation process.

2.2   SIGNATURES BASED ON COMPOSITION VECTORS

Clearly, it is possible to define many different types of composition vectors using different functions of oligonucleotide content. Moreover, several of them exhibit significant species specificity while being sufficiently conserved within a genome. We will briefly review a subset of composition vectors which represent an over- and underabundance of oligonucleotide usage or short-term dependencies in DNA.

2.2.1   MARKOV MODELS

Markov models have been frequently used to detect intragenomic heterogeneities. Models trained on coding and noncoding sequences were employed to predict gene sequences from open reading frames [Salzberg et al., 1998]. Different evolutionary pressures create compositional differences in genes and intergenic regions in an intragenomic scale, and Markov models are able to distinguish between the compositional differences of coding and noncoding regions. Intuitively, we can expect Markov models to capture the global compositional features in intergenomic scale. This was noted by Salzberg and colleagues [Delcher et al., 2007], who used their variable-order Markov-model-based gene prediction program for the classification of genomic sequences from different organisms.

As genomic signatures, we can view Markov models as being the conditional probability of a base given its finite length context. Here, DNA sequences are assumed to be stationary random processes; and the probability of emitting a base is independent of the bases located outside the context of that base, i.e., the process has finite memory. Thus, the conditional probability is equal to:

$$p(x_i|x_{i-1}x_{i-2}\ldots) = p(x_i|L_i) \tag{2.11}$$

where L_i is the context of the base x_i. While this context can be a different length for different bases, as in variable-order Markov models, fixing the length of L_i generates fixed-order Markov models. In fixed-order Markov models, the model parameter can be estimated as the ratio of two different-sized oligomer counts:

$$p(x_i|x_{i-1}x_{i-2}\ldots) = p(x_i|x_{i-1}x_{i-2}\ldots x_{i-k})$$
$$= \frac{p(x_{i-k}x_{i-k+1}\ldots x_{i-1}x_i)}{p(x_{i-k}x_{i-k+1}\ldots x_{i-1})} \qquad (2.12)$$

For a kth order Markov model, every base has 4^k different contexts; therefore, the profile of 4^{k+1} different parameters can form a genomic signature. For the same context, the probabilities of the 4 bases sum up to 1; therefore, the last one can be calculated from the other three. The genome signature profile, thus, has 3×4^k free parameters.

Dalevi et al. used Markov models as genomic signatures and determined that they are more specific than oligonucleotide counts [Dalevi et al., 2006]. They repeated their experiments using variable order Markov models and reported a slight improvement in species specificity with this modification.

In Figure 2.3, the multiple hypotheses true positive ratios are plotted for different short genomic fragment lengths. Comparing the Markov models with oligonucleotide frequencies of the same order, it can be seen that Markov models become more species specific for all oligonucleotide lengths.

Table 2.2: F-scores of one-way-ANOVA for Markov model parameters. Distribution of profiles for varying fragment length at different clade levels are considered.

F-value	1 kbp	10 kbp	50 kbp
Genus	312.88	2891.7	9079.99
Family	328.91	3041.5	9307
Order	364.82	3402.8	10004.2

2.2.2   ABUNDANCE PROFILES OF OLIGONUCLEOTIDES

The heavy-tailed behavior of k-distributions implies a significant over- and underabundance of oligomers within a genome. In fact, this is an indication of dependencies of nearby nucleotides, resulting in the deviation of their frequencies of occurrence from their expected values. We can expand Karlin's abundance measurement scheme based on Markov assumption to general k-mers.

Consider a k-mer $x_1, x_2, \ldots, x_k$. with probability $p(x_1, x_2, \ldots, x_k)$. We can write this probability as:

$$p(x_1, x_2, \ldots, x_k) = p(x_k|x_1, x_2, \ldots, x_{k-1})p(x_1, x_2, \ldots, x_{k-1}) \qquad (2.13)$$

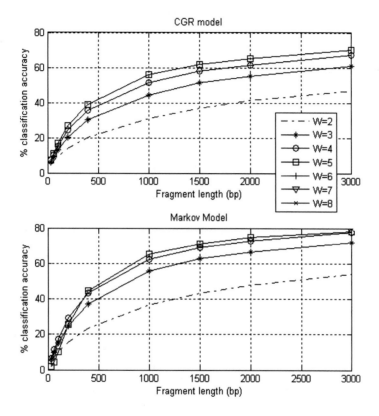

Figure 2.3: The multiple hypotheses true positive ratios for different short genomic fragment lengths. CGR model (i.e., frequency counts with Euclidean distance metric) vs Markov models determined using the same oligonucleotide length.

We can rewrite the first factor on the right-hand side of Equation (2.13) under different independence assumptions as follows. Assuming that the bases occur independently of each other, the conditional probability can be replaced by the marginal probability:

$$p(x_k | x_1, x_2, \ldots, x_{k-1}) = p(x_k) \tag{2.14}$$

Now we can calculate the odds ratio of an oligonucleotide frequency and its expected value based on this zeroth-order Markov model:

$$cv_0(x_1, x_2, \ldots, x_k) = \frac{p(x_1, x_2, \ldots, x_k)}{p(x_k)p(x_1, x_2, \ldots, x_{k-1})} \tag{2.15}$$

If we assume that the bases follow a first order Markov model,

$$p(x_k|x_1, x_2, \ldots, x_{k-1}) = p(x_k|x_{k-1}) \tag{2.16}$$
$$= \frac{p(x_{k-1}, x_k)}{p(x_{k-1})} \tag{2.17}$$

The corresponding relative abundance index is then given by:

$$cv_1(x_1, x_2, \ldots, x_k) = \frac{p(x_1, x_2, \ldots, x_k)p(x_{k-1})}{p(x_{k-1}, x_k)p(x_1, x_2, \ldots, x_{k-1})} \tag{2.18}$$

If the particular k-mer occurs more frequently than would be predicted based on the first-order Markov model, $cv_1(x_1, x_2, \ldots, x_k)$ will be positive, otherwise it will be negative; the magnitude will depend on how far the actual distribution of the oligomer varies from the prediction. Continuing in this fashion, we obtain:

$$cv_2(x_1, x_2, \ldots, x_k) = \frac{p(x_1, x_2, \ldots, x_k)p(x_{k-2}, x_{k-1})}{p(x_{k-2}, x_{k-1}, x_k)p(x_1, x_2, \ldots, x_{k-1})} \tag{2.19}$$

$$cv_3(x_1, x_2, \ldots, x_k) = \frac{p(x_1, x_2, \ldots, x_k)p(x_{k-3}, x_{k-2}, x_{k-1})}{p(x_{k-3}, x_{k-2}, x_{k-1}, x_k)p(x_1, x_2, \ldots, x_{k-1})} \tag{2.20}$$

$$\vdots \quad \vdots$$

$$cv_{k-2}(x_1, x_2, \ldots, x_k) = \frac{p(x_1, \ldots x_k)p(x_2 \ldots x_{k-1})}{p(x_2, \ldots x_k)p(x_1, x_2, \ldots x_{k-1})} \tag{2.21}$$

Therefore, a general scheme for calculating the deviation of oligonucleotide frequencies based on Markov models of order i $(i < (k-1))$ can be defined.

Table 2.3: (1): F-scores of one-way-ANOVA for Zeroth order Markov model profiles. Distribution of profiles for varying fragment length at different clade levels are considered.		
F-value	1 kbp	10 kbp	50 kbp
Genus	244.81	2515	7192.9
Family	254.81	2574.4	7903.4
Order	255.5	2713	7569.3

Table 2.4: (1): F-scores of one-way-ANOVA for Zeroth order Markov model profiles. Distribution of profiles for varying fragment length at different clade levels are considered.

F-value	1 kbp	10 kbp	50 kbp
Genus	135.59	1911	5274.4
Family	143.88	2042.3	6208.6
Order	160.67	2326.4	6513.9

Table 2.5: (3): F-scores of one-way-ANOVA for cv_{k-2} profiles. Distribution of profiles for varying fragment length at different clade levels are considered.

F-value	1 kbp	10 kbp	50 kbp
Genus	48.6	819.53	2327.5
Family	54.7	932.37	2517.4
Order	64.69	1102.9	2977.8

2.3 OLIGONUCLEOTIDE FREQUENCY DERIVED ERROR GRADIENT (OFDEG)

The signatures we have covered to this point are related with either the short-term or the medium-term dependencies of DNA sequences; and they are expressed as profiles, so they are elements of a multidimensional space. Oligonucleotide frequency derived error gradient (OFDEG), on the other hand, is a recently defined genome signature calculated based on the convergence rate of oligonucleotide frequencies estimated with increasing sequence length [Saeed and Halgamuge, 2009]. The biological foundation of this signature has not been explored yet; however, in practice, OFDEG is observed to be very species specific and pervasive although it is represented with only one parameter.

The oligonucleotide frequencies in a genomic fragment are clearly a better estimate of the oligonucleotide content than the estimate gathered from a subsequence of this fragment; because according to the ergodicity assumption, as the number of samples (i.e., longer fragment) increases, the estimates converge asymptotically. OFDEG simply attempts to capture this convergence behavior by subsampling the fragment and measuring the decrease in error as the length of the subsamples increases up to the fragment length.

The derivation of OFDEG is as follows. For a given fragment, the oligonucleotide frequencies of length k are calculated and stored in the OF_{full} vector. Starting with an initial subsequence of length L_1, random p subsequences are drawn from the given fragment; and the oligonucleotide frequency is calculated over each subfragment. The errors in the frequency counts are stored as:

$$e_{1,j} = ||OF_{full} - OF_{L_1,j}||, \qquad (2.22)$$

where $j \in \{1, 2, \ldots, p\}$. Increasing the subfragment length by l bp, p subsequences are sampled at each iteration; and the errors are calculated in the same fashion:

$$e_{i,j} = ||OF_{full} - OF_{L_1+il,j}||, \qquad (2.23)$$

where $i \in \{1, 2, \ldots, n\}$ and n is the last iteration number determined by subfragment length reaching some percentage of the original fragment (typically 80%). The relation of increasing subfragment length and decreasing error in correspondence is observed to be a linear decay. The last step of the OFDEG calculation is the measurement of the gradient of this decay by linear regression. The slope of the regression line gives the characteristics of the genome and is used as a genome signature. In Figure 2.4, the relationship is plotted for different genomes where the species specificity of the decay gradient can be observed. For a comprehensive set of prokaryotes, using multiple hypothesis testing by classifications, it was observed that the true positive detection ratios of OFDEG derived from tetranucleotide frequencies are comparable to the specificity of tetranucleotide frequency vectors for genomic fragments in the range of 8 bkp.

Table 2.6: F-scores of one-way-ANOVA for OFDEG values. Distribution of profiles for varying fragment length at different clade levels are considered.			
F-value	1 kbp	10 kbp	50 kbp
Genus	133.34	679.88	1740.5
Family	137.11	699.01	2034.1
Order	161.76	825.29	2442.9

2.4   DNA ENTROPY

The compositional complexity of a DNA sequence has been of interest in computational biology both for its evolutionary implications and for the application tools developed based on the entropy estimations of DNA sequences. Entropy has not been defined as a genomic signature, yet close connections with other signatures can be observed. Entropy is a measure of uncertainty in a sequence or a random process. It provides an indication of the complexity of sequences. The information embedded in biological sequences can be highly organized, and structured via high level grammatical

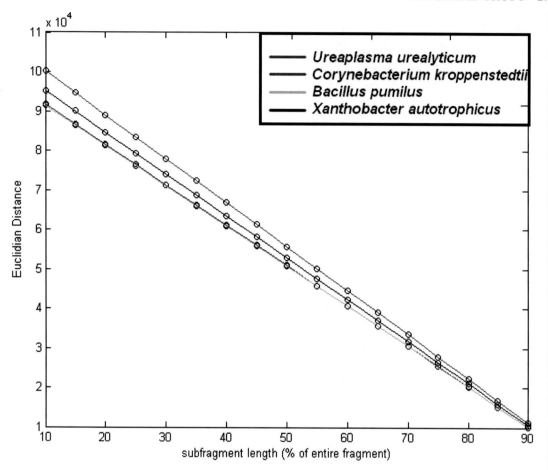

Figure 2.4: Signature error vs fragment length for four eukaryotic genomes (U. urealyticum, C. kroppenstedtii, B. pumilus, X. autotropicus). It can be seen that different species attain different error decay slope values.

rules. The currently available statistical and information-theoretic tools can only capture the structure in a low level organization. Nevertheless, even the information captured at a low level can be beneficial for many practical concerns as well as for investigating the evolutionary processes undergone by the genomes.

2.4.1   METHODS OF ESTIMATING DNA ENTROPY

Estimating DNA entropy is a great challenge, and a number of methods have been proposed. The most straightforward approach is using the block entropies of oligonucleotide content. One estimator of entropy utilizing word frequencies that is worth mentioning is the Renyi estimator [Vinga and Almeida, 2004, 2007] which interpolates the probability distributions with kernel-based methods (e.g., Parzen windows). However, the most efficient way to estimate the complexity of a genomic sequence is using specialized methods called DNA compressors. Since, a sequence cannot be represented at a lower rate than its per symbol entropy [Shannon, 1948], the size of the compressed fragment can be used as an upper bound of the entropy.

2.4.1.1 DNA Compression

Well-known compression programs working on finite alphabet sequences based on Lempel-Ziv (LZ) coding [Ziv and Lempel, 1977, 1978] or predictive coding [Sayood, 2005] are used as the estimators of complexity in biological sequences. However, these general purpose coders are not designed for special types of data, such as DNA, and therefore, DNA sequences are compressed inefficiently by conventional methods. Special purpose compression algorithms are designed and used for estimating the DNA entropy.

Probably one of the most well-known DNA compressors is Gencompress [Chen et al., 2000] which is motivated by the fact that DNA sequences contain tandem repeats, multiple copies of genes and palindromic sequences. It involves a modified LZ algorithm which searches for reverse complements and approximate repeats. The approximate repeats are the ones which have a small number of edit operations (substitution, insertion, deletion) in addition to a complete repeat. Copying an approximate repeat and modifying it with edit operations is shown to be cheaper than representing the original sequence. Gencompress performs better on average compared to the previous well-known programs, Biocompress and Biocompress-2 [Grumbach, and Tahi, 1993], which only consider exact repeats and short range correlations by employing a second order arithmetic coder in nonrepeat regions. Gencompress was later improved by DNAcompress which has better approximate search modules.

[Matsumoto et al., 2000] used a context tree weighting model which performs better with an additional LZ component to capture approximate repeats. It has also been shown that the algorithm compresses protein sequences. Cfact [Rivals et al., 1995] is another popular method which compresses the biological data in two passes. In the first pass, a suffix tree is trained and then encoded. [Behzadi and Fessant, 2005] find repeats to the cost of a dynamic programming search and select from a second order Markov model, a context tree and 2-bit coding for the nonrepeated parts.

To date, the best compression ratios are reported by [Cao et al., 2007]. They use an expert model with Bayesian averaging over a second order Markov model, a first order Markov model estimated on short-term data (last 512 symbols) and a repeat model. Weighting of probabilities for each model are based on the minimum description length (MDL) of the corresponding model. One advantage of this method is that it assigns probabilities to each symbol to be encoded; therefore, we can evaluate the information content of each spatial region. Expressed per element of complexity, this method can give an idea of the structure of the regions and the local properties of a genome.

2.4.1.2 Sufficient Statistics For Entopy Estimation

Although using compression programs for finding the upper bound of DNA entropy is practical, instead of measuring the encoded length of a sequence, direct estimation of entropy might be preferable. This would make it possible to eliminate the extra uncertainty due to the internal dynamics of a compression algorithm. For instance, compression algorithms must be causal because the main purpose is to transmit the sequence in minimum number of bits to a decoder which has no access to non-transmitted data. Thus, the algorithms encoding the data are on-line. While this is in accordance with how temporal data such as audio or speech are generated, biological sequences are spatial strings without sense of direction. A nucleic acid is dependent on the residues placed to the right of it as well as the ones to the left of it since both sides are involved in the composition. [Bejerano and Yona, 2001] trained suffix trees in both directions of biological sequences and reported no difference, indicating there is no right or left causality although it had been speculated that translation direction might affect the folding patterns in amino acid chains. They, then use two-way prediction which is a non-causal model.

Estimating the entropy without compressing the sequences finds room for applications measuring distances between DNA sequences. Average common substring which is the average length of longest substring matches between two sequences is used as the sufficient statistics for entropy estimation [Ulitsky et al., 2006]. It is proven that the entropy of a stationary ergodic source is asymptotically:

$$h = \lim_{N \to \infty} \frac{\log(N)}{\langle L(w) \rangle} \tag{2.24}$$

Where N is the sequence length and $\langle L(w) \rangle$ is the average word length of an LZ parsing [Grassberger, 1989, Cover and Thomas, 2006].

2.4.2 HISTORICAL NOTES ON BIOLOGICAL COMPLEXITY AND DNA ENTROPY

While an estimate of the entropy of genomes would provide information about the genotype complexity, its link to phenotype complexity and consequently the organization of life is not a new source of curiosity. Gatlin [Gatlin, 1963, 1966, 1968] searched for a link of phylogeny, life complexity and genomic entropy in the 1960's. Gatlin proposed a definition for the information content of DNA which was essentially a measure of the divergence of the DNA sequence from an *iid sequence*. Given an alphabet of size N, where N is four for DNA sequences and 20 for amino acid sequences, Gatlin defined two quantities D_1 and D_2 which measured divergence from the equiprobable state and divergence from independence, respectively.

$$D_1 = \log N - H_1(X) \tag{2.25}$$
$$D_2 = H_1(X) - H(X|Y) \tag{2.26}$$

where $H_1(X)$ is the first order entropy of the sequence and $H(X|Y)$ is the conditional entropy where X and Y are neighboring elements in the sequence. The information content of a DNA sequence is then defined as the sum of these two measures of divergence which can be shown to be the difference between the maximal entropy log N and the conditional entropy $H(X|Y)$ [Shannon, 1948]. Gatlin connects this definition of information to redundancy by noting that defining redundancy as

$$R = 1 - \frac{H(X|Y)}{\log N}$$

results in

$$R \log N = D_1 + D_2.$$

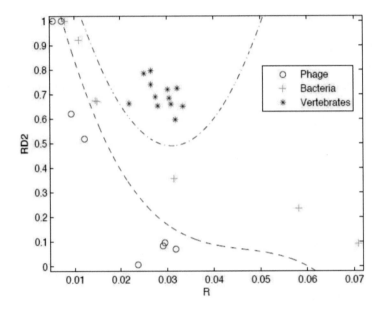

Figure 2.5: Plot of the redundancy rate versus $\frac{D_2}{D_1+D_2}$ using the genes available at the time shows a clear segregation of Phage, Bacteria, and Vertebrate sequences.

Based on the limited data available at that time, Gatlin showed empirically that DNA from vertebrates, bacteria, and phage (viruses that prey on bacteria) can be distinguished by looking at their information content and that there is increasing redundancy in the DNA of organisms as we move from lower complexity organisms like bacteria to higher complexity organisms such as vertebrates. Plotting the data available to her as shown in Figure 2.5 it is easy to see why she would come to that conclusion.

However, if we add more sequences from these groups the clear segregation breaks down as shown in Figure 2.6.

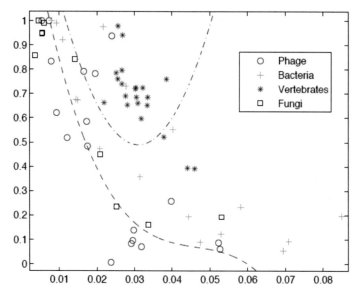

Figure 2.6: Inclusion of additional sequences breaks down the segregation observed by Gatlin.

2.4.3   CORRELATION OF DNA ENTROPY WITH OFDEG SIGNATURES

We have reviewed the OFDEG signatures, belonging to a distinct class of signatures which neither use the short term correlations directly, nor utilize the mid- or long-range correlations as in the case of previously introduced genomic signatures. Here, we show that this signature is connected to the complexity of genomes. By complexity, we refer to an upper bound of genome entropy obtained by oligonucleotide distributions. If the distribution of words of length k throughout a genome is close to random, the block entropy of the genome will be high. That is because, the uniform distribution constitutes the maximum uncertainty case for discrete distributions and the more even the distribution of oligomers in a DNA sequence, the higher its block entropy gets. On the other hand, a highly biased distribution of oligomers (i.e. exhibiting heavy tail behavior in k-distributions, having extreme overabundance and underrepresentation of certain words) results in lower entropy.

A similar relationship exists for OFDEG signatures. Considering that OFDEG measures the convergence of oligonucleotide frequency estimation with increasing genomic fragment length, the convergence rate is a function of genomic complexity. As two extreme cases, we can consider the usage of only one word in a sequence (e.g., a homopolymer) and the equal usage of each word homogeneously (e.g., a random sequence with IID uniform distribution). In the former extreme, the entropy of the sequence is minimum. Now consider the OFDEG value it attains. Since only one entry of the oligonucleotide count is incremented, the decrease in the error will be the same with the subfragment length subtracted from the total fragment length. Consequently, the gradient will be equal to -1, which is the minimum possible. On the other hand, for the other extreme the block entropy estimation will be maximum because the distribution appears to be uniform. The oligonucleotide count will be equally distributed for each word count and the expected decrease in error will be minimum with an expected slope of

$$-\sqrt{\sum_{4^k}(\frac{L}{4^k})^2/L} = -2^{-k}. \tag{2.27}$$

This is the minimum magnitude of an OFDEG signature parameter. In the midrange, the monotonic relation of the block entropy and the OFDEG is observed. And empirically, this relation is almost linear. To see that for the simplest case, we can observe the behavior for $k = 1$ which is actually the OFDEG derived from the GC content, and the entropy of the GC content. With the homongenity assumption in theory, the function of OFDEG with varying GC content ρ and the GC entropy are:

$$OFDEG_{GC}(\rho) = -\sqrt{\rho^2 + (1 - \rho)^2} \tag{2.28}$$
$$H_{GC}(\rho) = -\rho \log_2(\rho) - (1 - \rho) \log_2(1 - \rho).$$

This theoretical relation and the observed relation in actual genomes can be seen in 2.7. An almost linear relationship can be seen from the figure.

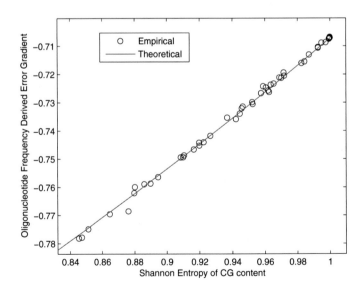

Figure 2.7: The Shannon entropy of the GC content of genomes vs the corresponding OFDEG values. The red line represents the theoretical relationship derived in Equation 2.28 and the circles represent the actual values.

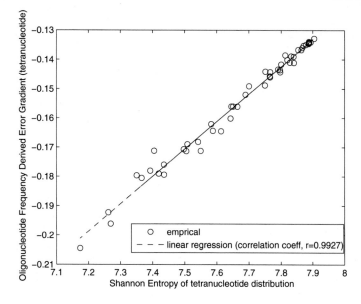

Figure 2.8: The Shannon entropy of microbial genomes derived from tetranucleotide frequencies vs the corresponding OFDEG values. A high correlation ($r^2 = 0.9927$, $p < 0.001$) can be seen from the plot.

We also compare the OFDEG values gathered from the prokaryotic genomes, and related entropies using linear regression. Tetranucleotide frequencies are used for the derivation of the signatures. A high correlation ($r^2 = 0.9927$, $p < 0.001$) empirically shows the tight connections of OFDEG signatures, genome complexity and entropy (figure 2.8).

CHAPTER 3

Measuring Distance of Biological Sequences Using Genome Signatures

In the previous chapters, we have viewed computational genomic signatures as mathematical structures mapped from DNA sequences to a metric space. These structures summarize the intrinsic dynamics of genomes and contain signals emphasizing the net response of the genome to environmental conditions. Throughout the discussion of computational genomic signatures, we have focused on the two basic signature features: specificity and pervasiveness. The former determines the distinguishability of different genomes, and the latter determines its usefulness when only fragmentary information about the genome is available. An absolute quality measurement for genome signatures is hard to define. However, using fundamental statistical tests on the signatures sampled from existing genomes helps us to compare signatures and determine their relative power. We have observed that different characterizations of DNA leading to different signatures vary in relative quality. This can be attributed to the capability of a signature to emphasize specific signals and capture native structural properties, as well as the homogeneity of the captured features. For example, as the length of the oligonucleotide increases, the specificity of signatures based on the oligonucleotide counts increases. This is because a k-mer profile already includes the information gathered by an l-mer profile ($l < k$) as the l-mer profile is a linear projection of the k-mers.

However, the structure of signatures is not the only factor determining their quality. How we interpret the signatures quantitatively also effects their specificity and pervasiveness. Different distance measures can affect the utility and, therefore, the power of a signature. It may be possible to better differentiate genomes with the same signature depending on how we measure distances between signatures. In this sense, the mathematical characterization of DNA sequences and the metrics proposed to compare these characterizations are both components of genome signatures. We have denoted the mathematical characterization as the signatures; because with distances other than standard norms, the signature becomes implicit. The defined metric maps the characterizations from the Euclidian space to another metric space and calculates the distance in the mapped space. However, since this mapping is not necessarily explicit, we cannot have a representation of the corresponding signature; and thus we stick with the definition in two parts, *genomic signature + distance measurement*, as the total characterization.

To exemplify the importance of the distance measure, consider the oligonucleotide frequency signatures with two different distance metrics. A thousand random genomic fragments from 99 prokaryotic genomes (each one picked from a different genus) were sampled for different fragment lengths. The signature used to represent each fragment was the vector of penta-nucleotide frequencies. It is expected that signatures of fragments from the same genomes are similar to each other, and signatures of fragments from different genomes are different from each other. Therefore, we expect the signatures of fragments from the same genome to be clustered together in the composition space. As a result of this clustering, it is possible to classify these signatures using supervised classification algorithms. We chose maximum margin classifiers, and performed ten-fold cross-validation and measured the ratio of true positives, which is a measure of the quality of the signature. The results were obtained by repeating the test with two similarity measures. First, we measured the similarity (S_1) of two signatures as the dot product of 5-mer vectors, where f_i and f_j represent the pentamer frequency vectors for genome fragments i and j. Second, the similarity S_2 was obtained via the Gaussian Kernel. The two distance metrics are:

$$S_1(f_i, f_j) = f_i^T f_j \tag{3.1}$$
$$S_2(f_i, f_j) = exp(-\frac{\| f_i - f_j \|}{\sigma})$$

It can be shown that using the Gaussian kernel in this manner is equivalent to mapping the input vector into an infinite dimensional space and taking the inner product in that space [Schoelkopf and Smola, 2002]. That is:

$$S_2(f_i, f_j) = exp(-\frac{\| f_i - f_j \|}{\sigma}) = \Phi_\infty(f_i)^T \Phi_\infty(f_j). \tag{3.2}$$

The results with the two similarity measures are shown in Table 3.1 There is a significant difference in the accuracy of classification. The species specificity and pervasiveness obtained using the Gaussian kernel is clearly superior. We can view this result in two different ways. Because of the kernel duality, the Gaussian similarity metric can be assumed to be the inner product of the signatures, $\Phi_\infty(f_i)$ and $\Phi_\infty(f_j)$. The superiority is because the implicit signatures, $\Phi_\infty(f_i)$ and $\Phi_\infty(f_j)$, capture the characteristics of the sequence better; or the improvement can be attributed to the similarity measurement and because the kernel similarity exploits the signals in the same genome signature better than the dot product calculation. In either case, the importance of the distance measure is evident.

In this chapter, we will review a number of difference/similarity measurement schemes in the context of particular signatures. The total operation can be interpreted as implicit signatures with better characterization of genomes; or it can be interpreted as the use of measures exploiting the information embedded in the same signature better than the conventional measures. Moreover, these measures are particularly important for some classes of genomic signatures (e.g., genomic entropy) which carry significant information about the specific and conserved structures of genomes but cannot be exploited using different measures.

Table 3.1: The 10-fold cross validation results of 1000 genomic fragments gathered from 99 genera for varying fragment length. The % true positive ratios are supplied.

Similarity metric	400 bp	1000 bp	2000 bp	5000 bp
S_1	51.3%	71.1%	75.3%	82.9%
S_2	69.5%	81.1%	87.5%	91.8%

3.1 CLASSICAL METHODS: EUCLIDIAN DISTANCES AND CORRELATION STATISTICS

The most popular sequence similarity/distance measurement for genome signatures are based on the simple ℓ-norms or correlations of signature profiles expressed as vectors in a multidimensional space. A well known example of this is the δ-distance measure of Karlin et al., which is a version of the ℓ_1 norm:

$$\ell_1(S_1, S_2) = \sum_i |S_1(i) - S_2(i)|, \qquad (3.3)$$

Karlin et al. used the δ-distance measure with genomic signatures based on the di-, tri- and tetranucleotide abundance vectors [Karlin et al., 1992, 1994a,b,c,d, 1995, 1997a, 1998]. The ℓ_1 norm was also utilized in the calculation of short-range and midrange correlation strength profiles [Dehnert et al., 2003, 2005a,b, 2006]. In order to quantify the differences and similarities between the CGRs of different sequences Deschevanne et al. defined the Euclidian distances of images calculated from pixel differences, which actually corresponds to the Euclidian distance of oligonucleotide frequency vectors [Deschevanne et al., 1999, 2000]

$$\ell_2(S_1, S_2) = \sum_i (S_1(i) - S_2(i))^2. \qquad (3.4)$$

A machine learning methodology based on unsupervised neural networks called self-organizing maps has been employed for the purpose of clustering short genomic fragments of the same origin together with the help of genomic signatures [Teeling, 2004, Abe et al., 2006, Chan et al., 2008a, Abe et al., 2005]. Self-organizing maps use Euclidian distance in the training of neurons, thus these methods can be considered to be in the class of genomic signatures used with Euclidian distances. OFDEG signatures [Saeed and Halgamuge, 2009] and tetranucleotide barcodes [Zhou et al., 2008] have also been used with the same metric.

Another popular technique to measure DNA sequence similarity is the Pearson correlation of genomic signature profiles:

$$\rho_{S_1, S_2} = \frac{\sum_i (S_1 - \overline{S_1(i)})(S_2(i) - \overline{S_2(i)})}{\sqrt{\sum_i (S_1 - \overline{S_1(i)})^2}\sqrt{\sum_i (S_2 - \overline{S_2(i)})^2}}. \qquad (3.5)$$

The Pearson coefficient has been used to calculate the similarities of abundance profiles of k-mers calculated over $(k - 1)^{th}$ order Markov model expectations [Chu, 2006, Chu and Li, 2009, Mrázek, 2009, Teeling, 2004], as well as of abundance profiles of k-mers calculated over zeroth order Markov model expectations [Bohlin et al., 2009a, Bohlin and Skjerve, 2009b, Bohlin et al., 2008].

Correlation measurements and ℓ-norms perform well with the general characteristics of genome signatures. However, they obscure pervasive signals by averaging out genome-wide total signals. This phenomenon can be observed better in short genomic fragments. Consider a genome with certain oligomers that are either over- or underrepresented. This characteristic is expected to be homogeneously represented within the genome. Yet, gathering statistics from a short genomic fragment is perhaps not sufficient to compile a full profile of word preferences. Assume a fragment of length 200 bp where the genome signature is determined to be 7-mer frequencies. The profile sampled from this genomic fragment will only represent 200 7-mers and they may be observed several times (i.e. <200 words will be observed). The total number of possible 7-mers are $4^7 = 16,384$; therefore, in this case only around 1% of the full profile is represented. Nevertheless, the 7-mers with nonzero frequency of occurrence are mostly from the set of overrepresented 7-mers which means there are detectable pervasive signals even with an insufficient number of samples. However, comparing all possible words in the distance/similarity calculation gives weight to the nonrepresented; and because there are many more nonrepresented oligomers, this makes detecting the pervasive signals due to the overrepresented oligomers more difficult. The classical comparison methods of ℓ-norms and correlation coefficients are metrics which take all 4^7 signature parameters into account. Clearly, an adaptive comparison metric that takes only represented words into account could have done better.

3.2   DISTANCES BASED ON MODEL FITNESS

We have briefly discussed an inherent weakness of the classical genome signature similarity/distance measurement approach. Now, we introduce some relative similarity measurement approaches based on model fitness which are potentially better at exploiting pervasive genome signature signals and at mitigating the problems that occur with the classical distance metrics. This is particularly true in applications where it is necessary to detect the genome of origin of short genomic fragments. In most applications, the use of absolute distance/similarity metrics limits the employment of relatively longer oligonucleotide counts. Since all 4^k k-mer frequencies are involved in those similarity measurements, good estimates of all of these k-mer frequencies is necessary. This requirement implies that overfitting should be avoided since overfitting would dramatically drop the detection accuracy. In order to prevent overfitting, more data points are needed in order to accurately estimate frequencies of occurrence (i.e., longer fragments); and the number of parameters to be estimated should preferably be kept small(i.e., shorter oligonucleotides where 4^k is equal to the number of different oligonucleotides). Not surprisingly, all these methods work best with 4-mers and 5-mers with sequences $\geq$ 1-3 Kbp. However, longer-range correlations exist in DNA sequences which we would like to exploit for characterizations even with short sequence reads.

3.2.1 LIKELIHOOD FUNCTIONS

The first class of similarity measures we discuss can be viewed as likelihood functions of oligonucleotide probabilities estimated from relative frequency counts. This class of similarity measures was first used by [Sandberg et al., 2001] for detecting the origin of species of short genomic fragments of unknown source. In this setting, given a genomic fragment, the probability of a genome being the origin of the corresponding fragment is calculated as $P(G_i|f)$ where G_i is the ith genome in a set of organisms and f is the genomic fragment. The genome resulting in the highest probability ($\arg max_i P(G_i|f)$) is determined to be the origin of this fragment. According to Bayes' Theorem:

$$P(G_i|f) = \frac{P(f|G_i)P(G_i)}{P(f)}. \tag{3.6}$$

If the prior probability of observing a genome is assumed to be equal for all organisms, the source genome is determined to be the genome G_i which would result in the highest probability of observing the fragment. In terms of genome signatures, this is simply the probability of emitting the genome fragment f, with the oligonucleotide probabilities estimated from genome i. Assuming independence of different oligonucleotides, this probability calculation turns out to be the multiplication of related oligomer probabilities:

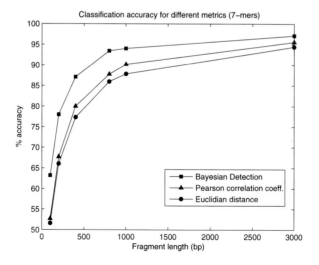

Figure 3.1: The accuracy performance of different distance/similarity metrics for 28 taxa with varying fragment lengths are shown. The frequencies of 7-mers are used. Using the metric defined by Sandberg et. al. appears to be more accurate for all fragment lengths than employing Euclidian distance and Pearson correlation coefficients.

$$P(f, G_i) = \prod_x P_{G_i}(x_1, x_2, \ldots, x_k)^{n_f(x_1, x_2, \ldots, x_k)} \tag{3.7}$$

where $P_{G_i}(x_1, x_2, \ldots, x_k)$ is the relative frequency of occurrence of the oligonucleotide $x_1 x_2 \ldots x_k$ computed from genome G_i, and $n_f(x_1, x_2, \ldots, x_k)$ is the number of times that oligonucleotide occurs in the fragment f. This probability estimate provides a measure of the likelihood that a fragment has been obtained from a particular genome based on the oligonucleotide content. Note that only the oligonucleotides observed in fragment f are involved in the calculation (i.e., nonobserved oligomers do not contribute to the product). This implies that the nonobserved oligomers are filtered resulting in the capture of pervasive signals and elimination of the noise stemming from the use of statistics of words not in the sample. Using this relative distance, Sandberg et al. were able to substantially reduce the size of the fragments that could be accurately classified obtaining a 90 percent classification accuracy for fragments of size 1.5 kbp in a set of 28 prokaryotic genomes from various genera. In Figure 3.1, the comparison of this measure with Pearson correlation and Euclidian distance is shown for 7-mer frequency signatures with various genomic fragment lengths. The use of a relative measure increases the quality of the signature making it more specific for all short fragment lengths. [Dalevi et al., 2006] extended the work of Sandberg et al. by replacing the probabilities in Equation 3.7 with conditional probabilities and variable-order Markov models. This turns out to be:

$$P(f, G_i) = \prod_x P_{G_i}(x_k | L_k)^{n_f(x_k | L_k)} \tag{3.8}$$

where L_k is the context of the kth base x_k determined by the Markov model trained on the genome G_i. L_k is the (k-1)-mer $x_1 x_2 \ldots x_{k-1}$ if the Markov model is of fixed order. Improvement over the likelihood function calculated by oligonucleotide content was reported [Dalevi et al., 2006], which is an improvement in the quality of the signature resulting from the change in profile (i.e., employing conditional probabilities instead of oligomer probabilities in the signature).

3.2.2   INDEXING BASED ON OLIGONUCLEOTIDE ABUNDANCE

The same idea of using relative measures which consider only observed words in a short genomic fragment can be extended to other signatures. In turn, signatures emphasizing the over- and under-abundance of oligonucleotides can be modeled in a profile and used as an index. Subsequently, the average scores attained by the oligonucleotides observed in a short genomic fragment can be used as a similarity measure. The abundance calculation for a k-mer can be obtained using an lth order Markov assumption ($l < k$). In this section, we describe such an indexing scheme which we call the relative abundance index (RAI).

In order to build a comprehensive abundance index it is useful if a combination of different order Markov models contribute to the characterization. We accomplish this in the following manner. First, we use models of various orders to predict the frequency of occurrence of the k-mer under consideration. We then use the log of the ratio of the observed frequency to the predicted frequency to provide an indication of how well or how poorly the k-mer follows the various Markov models.

Consider a k-mer $x_1, x_2, \ldots, x_k$. with probability $p(x_1, x_2, \ldots, x_k)$. We can write this probability as:

$$p(x_1, x_2, \ldots, x_k) = p(x_k | x_1, x_2, \ldots, x_{k-1}) p(x_1, x_2, \ldots, x_{k-1}) \tag{3.9}$$

We can rewrite the first factor on the right-hand side of Equation (3.9) under different independence assumptions. If the bases occur independently of each other, the conditional probability can be replaced by the marginal probability

$$p(x_k | x_1, x_2, \ldots, x_{k-1}) = p(x_k) \tag{3.10}$$

We can test this assumption by computing log-odds ratio as in Karlin et al. which we define as the relative abundance index of order 0.

$$rai_0(x_1, x_2, \ldots, x_k) = \log_2 \frac{p(x_1, x_2, \ldots, x_k)}{p(x_k) p(x_1, x_2, \ldots, x_{k-1})} \tag{3.11}$$

If we assume that the bases follow a first order Markov model, we can reduce the number of conditioning variables to one as:

$$p(x_k | x_1, x_2, \ldots, x_{k-1}) = p(x_k | x_{k-1}) \tag{3.12}$$

$$= \frac{p(x_{k-1}, x_k)}{p(x_{k-1})} \tag{3.13}$$

The corresponding relative abundance index rai_1 is then given by:

$$rai_1(x_1, x_2, \ldots, x_k) = \log_2 \frac{p(x_1, x_2, \ldots, x_k) p(x_{k-1})}{p(x_{k-1}, x_k) p(x_1, x_2, \ldots, x_{k-1})} \tag{3.14}$$

If the particular k-mer occurs more frequently than would be predicted based on the first-order Markov model, $rai_1(x_1, x_2, \ldots, x_k)$ will be positive, otherwise it will be negative. The magnitude will depend on how far the actual distribution of the oligomer varies from the prediction of the model. Continuing in this fashion we obtain:

$$rai_2(x_1, x_2, \ldots, x_k) = \log_2 \frac{p(x_1, x_2, \ldots, x_k) p(x_{k-2}, x_{k-1})}{p(x_{k-2}, x_{k-1}, x_k) p(x_1, x_2, \ldots, x_{k-1})} \tag{3.15}$$

$$rai_3(x_1, x_2, \ldots, x_k) = \log_2 \frac{p(x_1, x_2, \ldots, x_k) p(x_{k-3}, x_{k-2}, x_{k-1})}{p(x_{k-3}, x_{k-2}, x_{k-1}, x_k) p(x_1, x_2, \ldots, x_{k-1})} \tag{3.16}$$

$$\vdots \qquad \vdots$$

$$rai_{k-2}(x_1, x_2, \ldots, x_k) = \log_2 \frac{p(x_1, \ldots x_k) p(x_2 \ldots x_{k-1})}{p(x_2, \ldots x_k) p(x_1, x_2, \ldots x_{k-1})} \tag{3.17}$$

We can combine the RAIs of all orders by adding them to give:

$$rai(x_1, x_2, \ldots, x_k) = \sum_{i=0}^{k-2} rai_i(x_1, x_2, \ldots, x_k) \tag{3.18}$$

Given a particular k-mer $x_1, \ldots, x_k$, $\{rai(x_1, x_2, \ldots, x_k)\}$ gives an indication of how well the k-mer follows a Markov model. The smaller the model is that can predict the frequency of occurrence of the k-mer, the smaller will be the value of $\{rai(x_1, x_2, \ldots, x_k)\}$. For example, if the $k-$mer followed a third-order model but not a lower order model, one would expect the RAIs of an order greater than or equal to three to have a value close to zero. If the k-mer can only be explained by a fifth order model and not by a model of order less than five, then one would expect more of the coefficients to deviate from zero. In particular, k-mers that occur "unexpectedly" would have a high relative abundance index for all models and thus a high value in the sum of Equation (3.18). In this manner, $\{rai(x_1, x_2, \ldots, x_k)\}$ identifies oligomers that vary significantly from a set of Markov models.

The RAI profile provides index values for each oligonucleotide of length k; therefore the average score a fragment attains can be calculated as:

$$RAI(f, G_i) = 1/L \sum_{\mathbf{x}} n_f(x_1, x_2, \ldots, x_k) rai_{G_i}(x_1, x_2, \ldots, x_k). \tag{3.19}$$

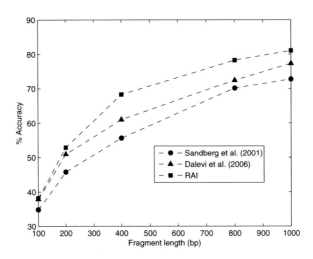

Figure 3.2: The comparison of an RAI measure with likelihood measures of oligonucleotide frequencies and Markov models for 100 bp-1000 bp fragment lengths. An oligomer length of 7 was used.

This index calculation also considers only the observed oligonucleotides in the fragment f since unobserved oligomers do not contribute to the sum. Therefore, this scheme emphasizes conserved words and provides a better representation of oligonucleotide abundance signatures in terms of pervasiveness and specificity. The corresponding classification performances are plotted in Figure 3.2.

3.2.3 MINIMUM DESCRIPTION LENGTH CALCULATION BASED ON LINGUISTIC MODELS

[Rissanen and Langdon, 1981] developed a general framework of hypothesis testing in which model selection is based on coding the data with models of different hypotheses and selecting the model which results in an encoding with a minimum number of bits. This idea is called the minimum description length (MDL) principle. We show that using this principle, it is possible to design efficient similarity measures for genomic signatures.

Given a random experiment and a set of possible outcomes $\{A_1, A_2, \ldots\}$ with probabilities of occurrence $\{p(A_i)\}$, [Shannon, 1948] showed that the number of bits required to encode the occurrence of outcome A_i is given as:

$$i(A_i) = -\log_2(p(A_i)) \qquad (3.20)$$

Thus, an event that occurs often will need fewer bits to represent it than an event that occurs less often. This will result in the minimization of the average number of bits needed to encode the outcomes of a sequence of experiments. The assumption here is that the distribution of outcomes used to determine the code is the same as the distribution of outcomes from the sequence of experiments. If the distribution of the outcomes from the sequence of experiments does not match the distribution used to design the code, then the number of bits used to encode or describe the outcomes may be quite high. We can view the code as a model and the average number of bits in the description, or the average description length, as a measure of how well the sequence of outcomes matches the model. We can build a number of codes, hence models, and attempt to identify the model that best describes a given set of observations. The codes themselves are not necessary; only the average description length corresponding to the codes, which can be estimated using the probabilities required to build the codes, is needed.

In the context of measuring the similarity of genomic fragments, we compute the description length for the fragment based on models generated by the target genomes. The genome corresponding to the model giving the minimum description length is assumed to have the same signature as the fragment and, hence, is determined to be the source of the fragment. As an example, employing fixed-order Markov models as the signature model, the description length estimation can be performed as follows. The number of bits required to express a base in a sequence is calculated using its conditional probability of that base occurring given several bases prior to the base under consideration in the $(5' \rightarrow 3')$ direction. This probability is estimated by the relative frequency counts observed from a given genome or an available portion of a genome. The number of bits required to represent the nth base in a sequence (x_n) given the genomic sequence G_i in the context of the k-mer upstream from that base, is:

$$I(x_n, G_i) = -\log_2(f_{G_i}(x_n | x_{n-1}, x_{n-2}, \ldots, x_{n-k}), \qquad (3.21)$$

where $x_{n-k} x_{n-k-1} \ldots x_{n-1}$ is the k-mer upstream from x_n. If the probability of occurrence of oligonucleotides in the genomic fragment F is given by $f_F(\mathbf{x})$, then the average number of bits required to encode the fragment using the model from genome G_i is:

$$MDL(F; G_i) = \sum_{\mathbf{x}} f_F(x_1, x_2, \ldots x_{k+1}) I(x_n, G_i). \tag{3.22}$$

Note that when the genomic signature is selected to be the fixed-order Markov model profiles, the minimum description length measurements become equivalent to the negative logarithm of the maximum likelihood functions introduced in Section 3.2.1. Here MDL calculates the total number of bits required to encode each base in the genomic fragment F determined by the coding model I. Replacing the relative frequency profiles obtained from the Markov models with another characterization scheme, a more general definition can be obtained as follows:

$$MDL(F; G_i) = \mathcal{M}_{G_i}(F), \tag{3.23}$$

where $\mathcal{M}_{G_i}$ is a coding scheme, trained on the genome G_i, and the fragment F is encoded using this scheme. The likelihood functions and index calculations are actually instances of this more general framework designed for oligonucleotide-usage-based genome signatures. Moreover, relative distance measurement by MDL enables embedding different characterization schemes as implicit signatures. A suitable example of such a model is dictionary coding. Efficient methodologies of sequence parsing for constructing dictionaries for given symbolic sequences were presented by [Ziv and Lempel, 1978]. We can apply dictionary coding to genomic sequences by scanning a genome in the $(5' \rightarrow 3')$ direction and progressively adding unobserved oligomers to the genomic dictionary. As the dictionary grows, existing oligomers in the dictionary become the context of new and longer oligomers due to the frequent occurrence of the corresponding words. This building block approach enables the capturing of the frequent usage of words regardless of their length. Therefore, LZ-dictionary building is useful to capture both short- and mid-term dependencies within a genomic sequence.

In the context of MDL, usage of LZ-dictionaries is straightforward: a given genomic fragment F is encoded using the dictionary built on the genome G_i, and the number of genomic dictionary entries used to represent the fragment, $\mathcal{M}_{G_i}(F)$, gives the minimum description length. If the fragment F is sampled from the same genome from which the LZ-dictionary was compiled, the MDL value tends to be small because similar word usage is expected in this genome globally due to the specificity and pervasiveness of the implicit genomic signature. An example with LZ78 algorithm was performed for the same dataset used for the experiment in Figure 3.2.

Table 3.2: The comparison of MDL measure (LZ78 dictionary coding is employed) with likelihood measures of Markov models (6th order Markov models are used) for 100 bp-1000 bp fragment length.

Similarity metric	100 bp	200 bp	400 bp	800 bp	1000 bp
$P(f, G_i)$	36.3%	51.2%	61.3%	69.5%	71.1%
$MDL_{LZ78}(f, G_i)$	39.7%	54.9%	73.4%	82.1%	86.3%

The results in Table 3.2 show that more specific genome signatures can be implicitly defined using MDL as the distance measure and dictionary generation methods as implicit genomic signatures when enough data for training is supplied. In that experiment, 90% of the genomes are used for training models; and the remaining 10% are used to sample genomic fragments.

CHAPTER 4

Applications: Phylogeny Construction

4.1   A HALF CENTURY OF MOLECULAR PHYLOGENETICS AND THE NEED FOR UNIVERSAL METHODS

Categorization of living organisms has always been of great interest, and continues to occupy the attention of scientists today. Starting with Aristotle, and perhaps with many prior unrecorded attempts, taxonomy has turned many corners, yet still remains a big challenge. It requires a hierarchical classification in a tree of life architecture, because all species evolve from a common ancestor. As we travel up the tree to the higher levels, the lower clade levels merge to form higher hierarchies. Defining the taxonomical levels coarsely, genera consist of species, further forming families. Families belong to the higher taxonomy of orders which form classes. All classes belong to a phylum. According to the currently accepted three-domain system [Woese et al., 1990], the tree of life consists of three main domains: *Archaea*, *Bacteria* and *Eukarya*. There are also intermediate group definitions between clade level pairs. Why we classify the tree of life is perhaps more of a philosophical question at its root. Nevertheless, in practice, systematics help science to achieve a better understanding of evolution and enable the search for the origin of the species. Furthermore, phylogenetic classification organizes the technical knowledge of medicine and food science [Gevers et al., 2005]. The challenge with taxonomy starts with how we define the categories and what they really mean, which are very highly debated issues [Rossello-Mora, 2005, Konstantinidis et al., 2006, Rossello-Mora and Amann, 2001, Stackebrandt et al., 2002, Staley, 2006]. Sometimes even defining unicellular organisms as different living units becomes a matter of discussion [Rossello-Mora, 2005]. Leaving this debate behind, another lack of consensus arises when the pigeonholing of species into taxonomical units is considered. The practical aim is to define methodologies for confirming the classification of currently identified species or to reclassify them, as well as to predict the location in the tree of life for newly discovered species. Currently, there is no officially accepted taxonomy, yet the nomenclature is governed with a degree of consensus.

Before the birth of molecular biology, phylogenetic classification was guided mainly by morphologic and phenotypic findings. An important cornerstone of systematics in the first half of the 20th century was the recognition of two main categories, prokaryotes and eukaryotes, by [Chatton, 1937]. These two main groups were differentiated by the existence of nucleated cells, and they also frequently differed in the techniques used to predict species' taxa. Phylogenetic knowledge and phenotypic coherence was the basis of eukaryote taxonomy for a long time since these features were relatively easy to observe. On the other hand, the prokaryote world, dominated by microscopic organisms, frequently required *in vitro* characterization, such as classification by cellular wall possession or by observing gram strains. The first migration to phylogenetic classification by microbiologists occurred only two decades after the discovery of DNA structure in the 1950s. Defining molecular criteria for groups of animals, plants, etc., took place later than microbes since replacing the conventional approach was not as urgent [Shneyer, 2007]. Since then, with the encouraging agreement between phenotypic observations and the genotype, molecular criteria based on biological sequence analysis have become the paradigm of taxonomy studies.

The gold standard of species definition for microbial organisms is recognized as the DNA-DNA hybridization technique [De Ley, 1970]. This technique began to be employed in systematics starting in the late 1960s. It defines two microbes as belonging to the same species if the ratio of molecular hybridization of DNA molecules exceeds a threshold. This threshold is around 70% [Wayne, 1987]; and, in most species, the hybridization was observed to be higher than this threshold [Rossello-Mora and Amann, 2001]. Although, DNA-DNA hybridization remains the gold standard for species definition, practical concerns and technical limitations lead the need for developing other criteria. This physical technique is time consuming, and it can only be used with cultivated microbes [Stackebrandt and Ebers, 2006]. Also its high error rates can result in misinterpretations. Lastly, it is good for defining species; but for higher clade levels, DNA-DNA hybridization fails to provide consistent results.

As a result of the drawbacks of DNA-DNA hybridization for high-level taxa resolution, a completely different technique was developed. This new paradigm consists of constructing phylogenies based on conserved marker regions in the DNA, and it is still the most well-known approach. The conserved markers used to represent species are generally picked from among gene sequences. However, the selection of marker genes should be done very carefully because of several problems. First of all, different genes are under different levels of evolutionary pressure, and, therefore, may not necessarily reflect the general characteristic of the organism. Attempts using metabolic enzyme genes failed as they gave different phylogenies for the same organism and even failed to construct the three domains of life correctly in some cases [Brown and Doolittle, 1997]. Second, due to frequent gene gain and loss events, the marker genes should be the elements of a universal pan-genome set so that an ortholog can be found for each species. Also, the genes may undergo homologous gene recombination even among divergent strains [Vulic et al., 1997]; and recombined genes introduce noise to the representations. Among the marker gene methods, perhaps the most popular one is using the small subunit of 16s rRNA genes [Olsen et al., 1994, Stackebrandt and Goebel, 1994, Ludwig et al., 1998, Woese and Fox, 1977]. Its popularity is because this 1500-1800 bp unit belongs to ribosomal RNA genes, which are conserved both temporally and spatially. That is to say, ribosomal RNAs are universal to microbial organisms; and 16s rRNA are not known to participate in lateral genetic transfer events. Also, compared to most classes of genes, mutations, insertions and deletions happen rarely in 16s rRNA small subunits. This conservative rate of change is believed to have approximately proportional synchronization with molecular clocks.

The phylogenetic distance of two species is estimated by alignment scores when 16s rRNAs are used as taxonomic markers. The species cut-off level of 70% hybridization corresponds to 94% similarity of 16s rRNA SSU [Goris et al., 2007]. The phylogenetic trees, constructed by 16s rRNA sequences, can define the prokaryotic clade levels accurately. It is reported that 16s rRNA identification mostly agrees with the reevaluated prokaryotic taxonomy of an ad hoc committee [Stackebrandt et al., 2002, Wang et al., 2007]. The employment of 16s rRNAs greatly mitigated the problem of DNA-DNA hybridization with ambiguous clade levels higher than species. In contrast, it introduced a complementary problem of intraspecies classification; because the strains of a species have very similar marker sequences [Stackebrandt and Goebel, 1994], even being identical in some cases (e.g., *S. aureus*). Hybrid usage of these two techniques is one approach to resolving this problem [Gevers et al., 2005]. Despite their practical use and accuracy, 16s rRNA methods also have other serious drawbacks. The diversity of microbes is thought to be very high, even possibly up to a billion species [Curtis and Sloan, 2004, Giovannoni and Stingl, 2005]. For this level of complexity, the resolution of 16s rRNA sequences is not sufficient, meaning that many species have to share very similar 16s rRNA sequences. In this case, the problem with strain identification will be projected to a global level.

A recent attempt to capture a better resolution with the marker gene approach is using a set of marker genes in combination. Basically, comparison of longer sequences means a higher number of substitution, insertion and deletion events and, thus, a higher resolution of sequence distances. A technique called multilocus sequence typing (MLST), using seven housekeeping genes as the conserved markers, which is able to distinguish organisms at the strain level has become increasingly popular [Hanage et al., 2005, Maiden et al., 1998].

Constructing eukaryotic phylogeny using molecular methods has also been a great challenge. Due to high heterogeneity and the complexity of eukaryotic genomes, DNA-DNA hybridization has been a poor identification technique in eukaryotes. The computational phylogeny studies on eukaryotes are more fragmented, and a number of diverse techniques focusing on different locations of genomes in different taxonomies have been proposed. After the development of PCR, investigating polymorphic DNA became popular; and this technique is still used in plant studies [Shneer, 1991, Bannikova, 2004]. Using 18s rRNA small subunits [Soltis et al., 1999] is also popular in plants. Mitochondrial DNA is found to exhibit a higher level of polymorphism than nuclear DNA in animals. It is also relatively more conserved and subject to less recombination. This stimulated the idea of using marker gene sequences from mitochondria, such as cytochrome b (cytb), and subunits of cytochrome c oxidase (COX-1 and COX-2), along with the ribosomal RNA genes. A number of other mitochondrial marker genes have also been used such as ribosomal RNA (small subunit 18s rRNA, large subunit 28s rRNA) from nuclear genomes, thought to be alternative marker genes in invertebrates [Floyd et al., 2002, Markmann and Tautz, 2005]. In vertebrates, the recombination activating gene 1 (RAG1) is another popular gene [Greenhalgh and Steiner, 1995]. In addition to genes, noncoding sequences from nuclear genomes and chloroplast genomes are also utilized as marker regions [Shneyer, 2007, Nickrent et al., 2004]. Usage of different markers in different eukaryotic phylogeny construction studies leads to consensus problems. To standardize the procedure, a barcoding scheme, which determined a good conserved 648 bp region from the 5'-end of the cytochrome c oxidase subunit 1 (COI) gene, was proposed [Hebert et al., 2003]. Using the similarity distances obtained using this marker, birds [Hebert et al., 2004], fish [Ward et al., 2005] and amphibians [Vences et al., 2005a] were classified. However, the use of barcodes as universal markers is problematic [Armstrong and Ball, 2005, Vences et al., 2005c], because factors like rearrangements in plants, uneven distribution of introns in fungi, etc., may make universal primers unavailable.

Although employing marker sequences for phylogeny construction is the currently operative paradigm, its problematic nature makes replacement with more universal methods inevitable. Most marker sequences have universality problems, because they are not conserved uniformly in the tree of life. Zooming into local clades, it is possible to find conserved sequences. But this tree-dependent optimization approach destroys universality and adjusts the resolution according to local needs [Gemeinholzer et al., 2006, Feliner and Rossello, 2007, Chase et al., 2005]. Moreover, the selection of markers is arbitrary to a level [Shneyer, 2007]; and the associated specific characteristics might not be correlated with the evolution of species. The evolution constraints are multidimensional; and only a part of these constraints apply to single genes or a collection of a few genes [Koonin et al., 2000]. It is generally accepted that for better capturing of evolutionary history, whole genome sequences should be included in the phylogenetic analysis. Initially, there were doubts as to whether comparing whole genomes would produce ambiguous phylogenies due to lateral gene transfers. However, it was shown that the effect of horizontally transferred genes was not enough to distort the phylogenetic signals [Eisen and Fraser, 2003]. Sequence distances defined on whole genomes, which are able to approximate evolutionary divergences, also eliminate the problem of nonuniversality because they are independent of locally conserved structures. The recent advances in DNA sequencing technology, with the increase in the number of completed whole genomes, enable the computational study of whole-genome phylogeny construction. Average nucleotide identity (ANI) [Konstantinidis and Tiedje, 2005], which is the pairwise alignment of complete ortholog gene sets of genomes, correlates well with DNA-DNA hybridization results [Richter and Rossello-Mora, 2009]. A similar relationship is also observed with whole-genome comparison with maximal unique matches [Deloger et al., 2009]. This shows the applicability of whole-genome analysis to taxonomy; however, the feasibility of these methods is an issue. Aligning prokaryotic genomes which have small length is achievable. Aligning long eukaryotic chromosomes produce enormous computational burdens which, in practice, are not feasible. An alignment-free alternative is the employment of genome signatures which carry signals of the evolution history of a species.

4.2   PHYLOGENETIC SIGNALS IN GENOME SIGNATURES

The ability to capture evolutionary trends has been the main theme of our discussion. We have seen that the average response of a genome to its environment leads to compositional differences. Carefully measuring these compositional patterns can provide decent species specificity. Therefore, observing phylogenetic signals in genome signatures is expected, but the question is, how do these signals function at different levels? If phylogenetic signals accumulate for long time periods without being corrupted, constructing phylogenies becomes possible using genome signatures. An interesting observation on the consistency of genome composition with species evolution can be obtained using the most simple genome signature, the GC content. In [Haywood-Farmer and Otto, 2003], a Brownian motion model is simulated with the parameters of mutation rates governing the GC content. The parameters were derived by a maximum likelihood estimation from molecular databases. It was observed that the divergence of GC content agrees with the evolutionary distances estimated from 16s rRNA sequence alignments. The GC content alone is not sufficient to construct phylogeny trees. Nevertheless, this indicates the potential of genome signatures for estimating evolutionary distances.

The consistency of dinucleotide abundance profile distances (calculated by the δ^*) distance), was shown by [Karlin et al., 1994b] for vertebrates and some fungi sequences according to the accepted orderings of these eukaryotes. For prokaryotes, the dinucleotide abundance signature supports some claims, such as classifying *Shigella spp.* and *Escherichia coli* as the same species [Pupo et al., 2000] or the similar situation of the *Bacillus cereus* cluster [van Passel et al., 2006]. Despite these findings, the dinucleotide abundance signatures are not sufficient to construct phylogeny trees on a global scale. Compared to some other prokaryotic phylogeny construction methods, dinucleotide signatures are considered with a resolution in the family-species range (Figure 4.1).

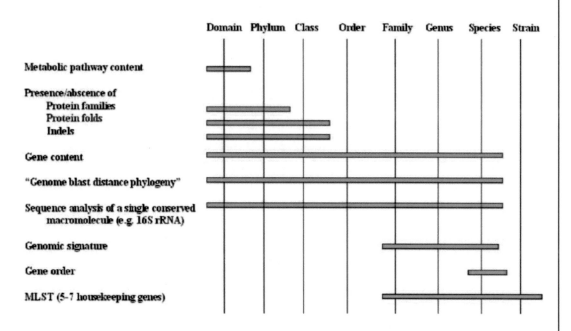

Figure 4.1: Different phylogeny construction methods and their resolutions in clade-levels. Figure taken from [Coenye et al., 2005].

A genome signature selection with higher species specificity can perform better than dinucleotide abundance ratios. Indeed, using tetranucleotide frequencies normalized by the zeroth-order Markov models consistent phylogenetic trees with 16s rRNA methods have been constructed [Pride et al., 2003]. However, there are some disagreements, such as *H. pylori* and *C. jejuni* do not share common ancestors with some members of the *Enterobacteriaceae* family. Noticeably, these organisms lack repair mechanisms which resulted in the rapid divergence of these organisms from their ancestors in composition [Parkhill et al., 2000, Tomb et al., 1997]. Tetranucleotide abundance signature distances are also reported to be consistent with DNA-DNA hybridization and whole-genome alignments [Richter and Rossello-Mora, 2009, Coenye and Vandamme, 2004].

Not surprisingly, beyond the oligonucleotide length of 2-4-mer, considering a longer context increased the accuracy of phylogeny construction using genome signatures. Employing the Euclidian metric as the distance measurement with oligonucleotide abundance content beyond tetranucleotides, the currently-accepted phylogeny trees up to the family level can be constructed within a range of tetra- to octanucleotides. Moreover, better resolution in intraspecies distances can be attained. This appears to be superior to 16s rRNA distances, which cannot resolve the strain-level phylogenies.

Pushing the limits further, better phylogeny construction has been reported for larger contexts. [Qi et al., 2004] used the abundance profiles comparing the observed oligomers with the expected values derived from $(k-2)^{th}$ order Markov models as:

$$a(\alpha_1, \alpha_2, \alpha_k) = \frac{p(\alpha_1, \alpha_2 \ldots \alpha_k) - p_e(\alpha_1, \alpha_2 \ldots \alpha_k)}{p_e(\alpha_1, \alpha_2 \ldots \alpha_k)} \tag{4.1}$$

where the expected frequencies are derived from (k-1)-mer and (k-2)-mers:

$$p_e(\alpha_1, \alpha_2, \alpha_k) = \frac{p(\alpha_1, \alpha_2 \ldots \alpha_{k-1}) p(\alpha_2, \alpha_3 \ldots \alpha_k)}{p(\alpha_2, \alpha_3 \ldots \alpha_{k-1})} \tag{4.2}$$

Qi et al. performed the calculations using oligopeptides (mainly for $k = 5$ and $k = 6$) instead of oligonucleotides. We associate this composition-based approach with genome signatures, because pentapeptide and hexapeptide frequencies can be approximately deduced from oligonucleotides with lengths of $k = 15$ and $k = 18$, respectively. The Pearson correlation subtracted from 1 was used as the distance measure. With a set of 109 genomes from all three domains of life, this composition vector method was able to separate all three domains of life successfully, as well as construct phylogenies similar to the 16s rRNA method for prokaryotes.

Later on, the compositional vector method and its variations were successfully applied to various tasks of phylogeny construction. [Stuart et al., 2002] calculated tripeptide and tetrapeptide frequencies from whole genomes of vertebrate mitochondria. Singular value decomposition helped for a dimension reduction from 20^4 ($or\, 20^3$) dimensional oligopeptide space to 38 dimensions of principal components. Their results agreed with the previously proposed phylogeny trees of primates, rodents, cetardiodactyls, carnivores, and perissodactyls [Xu et al., 1996, Janke et al., 1997]. In accordance with previously published results, close relationships between whales and hippos [Arnason et al., 2000], bats and ferungulates [Nikaido et al., 2000] and squirrels and other rodents [Reyes et al., 2000] were confirmed. The same dataset of vertebrate mitochondrial genomes was also compared using oligonu-cleotide abundance vectors of an oligonucleotide length of 11 calculated as in Equations 4.1 and 4.2. A topology similar to that described by Stuart et al. was obtained [Yu et al., 2010]. Whole genome analysis with k-mers was performed using chloroplast DNA. The construction of [Chu et al., 2004] was consistent with the currently accepted photosynthetic eukaryote taxonomy. The signature us-ing Equations 4.1 and 4.2 by Chu et al. was then replaced by a first-order Markov estimation of oligomers by [Yu et al., 2005] for the same dataset. In this case, the k-mer estimation is:

$$p_e(\alpha_1, \alpha_2, \alpha_k) = \frac{p(\alpha_1, \alpha_2 \ldots \alpha_{k-1})p(\alpha_k) + p(\alpha_1)p(\alpha_2, \alpha_3 \ldots \alpha_k)}{2}. \tag{4.3}$$

A selection of distances in which the Pearson correlation of abundance vectors subtracted from 1 gave trees converging to rRNA-based trees was proposed before [Charlebois et al., 2003]. A general trend for this class of signature-based genome phylogeny methods is a clear separation of β-Proteobacteria and ϵ-Proteobacteria where no clear segregation of β-Proteobacteria and γ-Proteobacteria groups is observed. In a recent note [Yu et al., 2010], the same group noticed that although the metric they used is practical, it does not satisfy the triangle inequality and, therefore, is not a proper metric. A modification normalizing the vectors to unit length, which satisfies the metric properties theoretically, also improved the phylogenetic trees inferred using compositional vectors.

4.2.1   PHYLOGENY WITH INFORMATION THEORETIC DISTANCE MEASURES AND IMPLICIT GENOME SIGNATURES

A general definition of information theoretic distance measurement based on Kolmogorov com-plexity

$$d(x, y) = \frac{\max\{K(x|y), K(y|x)\}}{\max\{K(x), K(y)\}} \tag{4.4}$$

enables approximate calculations by substituting practical complexity measurements for the theo-retical measure of algorithmic complexity. Equation 4.4 can thus provide alternative calculations of the information distance.

Using standard data compression programs, as well as the specialized DNA compression soft-ware, to approximate the genomic/proteomic sequence complexity has been a successful, straight-forward method. [Li et al., 2001] predicted the mammalian orders with the help of GenCompress run on mitochondrial DNA. Similar results were obtained using both DNACompress and com-pression using Prediction by Partial Match (PPM) [Dawy et al., 2005]. [Apostolico et al., 2006] use extensible motif compression for inferring phylogenies of Eutherian orders.

Implicit signatures imposed by entropy calculations were also used in a class of information theoretic sequence distance approaches for phylogeny constructions. Using k-mer frequencies to approximate the sequence entropy is a very practical approach. An example is the use of block entropies:

$$d_1(x, y) = \frac{\max\{D_{KL}(x||y), D_{KL}(y||x)\}}{\max\{H(x), h(y)\}} \qquad (4.5)$$

where $D_{KL}(.||.)$ denotes the Kullback-Leibler divergence (or relative entropy) and $H(.)$ stands for the block entropy calculated using k-mer frequencies.

One alternative estimating joint entropies and mutual information of DNA sequences using k-distributions of oligonucleotides, is performed as follows. The distance metric is defined as:

$$d_2(x, y) = 1 - \frac{I(x||y)}{H(x, y)}. \qquad (4.6)$$

In order to estimate the mutual information and the joint entropies, joint distributions of a pair of sequences have to be defined. The joint entropies are defined on k-distributions of oligonucleotides. Marginal and joint histograms are calculated as follows. $f_k(x) = (f_1, f_2, \ldots f_{4^k})$ are the oligonucleotide frequencies of k-mers for the genomic sequence x and $f_k(y) = (g_1, g_2, \ldots g_{4^k})$ are the k-mer frequencies for the sequence y. Assume x contains N different levels of oligomer frequency, and y has M different oligomer frequency counts. The k-distributions are:

$$p(x_i) = \tfrac{1}{4^k}\#\{l = 1, 2, \ldots, 4^k \ : f_l = x_i\}, i = 1, 2, \ldots, N, \qquad (4.7)$$
$$p(y_i) = \tfrac{1}{4^k}\#\{l = 1, 2, \ldots, 4^k \ : g_l = y_i\}, i = 1, 2, \ldots, M,$$
$$p(x_i, y_j) = \tfrac{1}{4^k}\#\{l = 1, 2, \ldots, 4^k \ : f_l = x_i, g_l = y_j\}, i = 1, 2, \ldots, N, j = 1, 2, \ldots, M.$$

The sequence distance turns out to be:

$$d_2(x, y) = 1 - \frac{\sum_{i,j} p(x_i, y_j) \log_2(p(x_i, y_j)/p(x_i)p(y_j))}{\sum_{i,j} p(x_i, y_j) \log_2(p(x_i, y_j))}. \qquad (4.8)$$

Consequently, the similarity of sequences is determined by the similarity of the tail behavior in oligonucleotide k-distributions. For the oligonucleotide length of k=12, 64 mitochondrial genomes were compared to infer phylogenies [Yu et al., 2007]. The topology of the inferred tree was reported to be roughly in accordance with the current taxonomy of vertebrates.

This upper bound on Kolmogorov complexity can be obtained by using block entropy estimations as in the previous two examples. It is also possible to approximate the entropy of sequences of unknown Markov sources from LZ dictionaries [Lempel and Ziv, 1976]. In this case the metric can be defined, by either:

$$d_3(x, y) = \frac{\max\{C_{LZ}(xy) - C_{LZ}(x), C_{LZ}(yx) - C_{LZ}(y)\}}{\max\{C_{LZ}(x), C_{LZ}(y)\}} \qquad (4.9)$$

or by

$$d_4(x, y) = \frac{\max\{LZ_x(y)LZ_y(x)\}}{\max\{LZ_x(x), LZ_y(y)\}}. \tag{4.10}$$

In d_3, $C_{LZ}(x)$ stands for the size of the LZ dictionary inferred from the sequence x; and xy stands for the concatenation of two genomic sequences. This approach was used by [Otu and Sayood, 2003] to infer the phylogenetic trees from mitochondrial genomes of Eutherian orders and was observed to be in agreement with previously known phylogenies. In d_4, $LZ_x(y)$ stands for the size of the encoded sequence of y, compressed by the LZ dictionary derived from the sequence x. This can be considered as a variation of the minimum description length approach since the length of the description of a sequence with a derived set of parameters governs the distance.

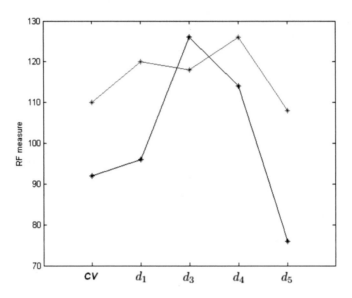

Figure 4.2: The RF distances of the phylogenetic trees inferred by composition vectors (CV), Kullback–Leibler divergence of oligomers (d_1), LZ complexity with sequence concatenation (d_3), LZ complexity with compression (d_4), and average common substring (d_5) methods. The trees are compared to 16s rRNA-based phylogenetic trees. Red lines=genome sequences, and black lines=proteome sequences.

A sequence distance measure called the Average Common Substring (ACS) approach [Ulitsky et al., 2006] can be categorized in LZ complexity-based methods. This scheme calculates the distance of two sequences as

$$d_5(x, y) = \frac{\log_2(|y|)}{L(x, y)} - 2\log_2(|x|)/|x|. \tag{4.11}$$

In the equation above, $|x|$ is the length of the sequence x; and $L(x, y)$ is the average longest substring match between sequences x and y

$$L(x, y) = \sum_i^{|x|} \ell(i)/|x|. \tag{4.12}$$

$\ell(i)$ is the length of the longest oligonucleotide match between x and y starting from i^{th} position of x. This is very similar to the LZ77 algorithm [Ziv and Lempel, 1977] in which the longest match is searched in the previous history of a sequence. As the search buffer, another sequence y is used in the case of ACS. Instead of actually coding the sequences, the average length of the matches is calculated as an approximation of the complexity. Clearly, d_5 is not a metric, since it does not support similarity. It is modified as $(d_5(x, y) + d_5(y, x))/2$. A set of complete genome and proteome sequences from all three domains of life are compared to construct the phylogenetic trees. In Figure 4.2.1, the Robinson-Fouldes measure for tree distances is shown for some of the distances introduced here. The phylogenetic trees inferred from whole genome and proteome sequences for the three domains of life are compared against 16s rRNA trees using the unshared interior branches as:

$$d_T(t_1, t_2) = 2(\min(p_1, p_2) - I(p_1, p_2) + |p_1 - p_2|) \tag{4.13}$$

where p_1 and p_2 are the number of partitions (inner branches) in the inferred trees, t_1 and t_2 and $I(p_1, p_2)$; gives the number of shared branches between them. It can be seen from Figure 4.2.1, that the composition vector method is also very competitive compared to the information theoretic measures, implying that an over- and underabundance of oligomers contains significant phylogenetic signals.

4.3   PHYLOGENY CONSTRUCTION BY GENOME SIGNATURES USING GENOMIC FRAGMENTS

Using genome signatures on complete genomes to infer phylogenies has been shown to be an effective and computationally feasible approach. The universality due to tree independence, which avoids the search for marker elements and sequence alignments, is perhaps the most powerful feature of this paradigm. The pervasiveness of currently known genome signatures is supported by a volume of literature. A practical concern is investigating the limits of phylogeny construction accuracy as often only short genomic sequences are available instead of complete genomes. If the genome signatures have this capability, this can lead to a practical solution of various contemporary problems in systematics (e.g., finding universal and feasible computational methods to construct the tree of life). To this day, a comprehensive set of studies on this topic has not been conducted. Fortunately, a few attempts employing the genome-signature-based phylogeny methods on marker sequences have been reported recently. In addition, positive findings on the phylogeny construction ability of genome signatures on partial genomic sequences encourages the claim.

The barcoding scheme used to catalogue vertebrates was implemented using the oligonucleotide frequencies, with an oligonucleotide length of 9-10-mers. Using the 648bp subunit of COI genes, it was observed that inferred phylogenies have superior resolution to sequence alignment [Chu and Li, 2009]. The composition vector method with 18s and 16s rRNA subunits was also able to perform accurate barcoding of vertebrates.

The taxonomic Naïve Bayesian Classifier offered by the Ribosomal Database Project (RDP) [Wang et al., 2007] is designed based on the octanucleotide signatures. The SeqMatch classifier, which is another taxonomical classifier offered by RDP, also uses the 7-mer usage. Both classifiers process 16s rRNA small subunits for taxonomy classification. Superior classification compared to sequence alignment is observed for both methods. Furthermore, even very short fragments, down to 200 bp, could provide satisfactory taxonomy classification of certain clade levels. Sequences around 400bp have been observed to resolve at the classification to the genus level, while 200 bp fragments can operate at the family level. According to these results, when applied on the marker sequences, genome signatures perform better than sequence alignments, although the markers are considerably well conserved with similar sequences. This superiority might imply that observing correlated changes and context-dependent mutations captures a better history of molecular evolution than the aligned point mutations.

An example of taxonomic classification with random genomic fragments was performed using correlation strength and average mutual information profiles. Capturing short-term correlations corresponding to the first 15 (or greater) components of the correlation strength vector, the exact trees for vertebrates could be inferred [Dehnert et al., 2005b]. This resolution can be preserved using genomic sequences of nuclear genomes as short as 25,000 kbp, which constitutes a small percentage of most vertebrate genomes.

In an investigation with normalized oligonucleotide frequencies with zeroth-order Markov models [Takahashi et al., 2009], it was reported that random genomic sequences as short as 3-7 kbp can resolve the phylogenetic trees to at least the family level. Even better resolution may be provided for longer fragments.

4.4 DRAWBACKS OF GENOME-SIGNATURE-BASED PHYLOGENY CONSTRUCTION

Computational genome signatures are able to offer universal, robust evolutionary distance measurement. These techniques are computationally feasible; and they do not require any location of specific markers or extensive procedures, such as sequence alignment. However, significant disagreements with known phylogenies and signature-based constructions are not rare. Initially, dinucleotide signatures were thought to be resolving in local scales that cannot provide significant information for higher taxa. Later on, signatures involving longer oligonucleotide content, which contain midrange correlation information (e.g., information theoretic distance measurements, AMI profiles, correlation strength profiles, etc.), were reported to resolve all three domains of life. Nevertheless, processing whole genomes brings in additional signals which corrupt precise phylogenetic signals. For instance, regions and genes evolving at different molecular clock speed might govern the signatures, on aver-age corrupting the related signals. Even though horizontal gene transfers alone are not sufficient to dominate the signatures, they also introduce noise to genome signatures.

A concern regarding genome signatures involves their resolution capacity. Most compu-tational genome signatures consist of a large number of parameters, which in theory should support a decent resolution. On the other hand, it can be shown that only a small volume of the corresponding abstract space is occupied by the species. Principal component analysis of a large class of genome signatures shows that most of the variation can be summarized with a few principal components [Mrázek, 2009]. The genome composition is driven by a few trends, and the actual rank of characterization is smaller than the number of observed parameters. Con-sequently, a crowded subset of the signature space could be occupied by millions or billions of species.

Comparing the distances derived from 16s rRNA sequence alignments and genome signatures from whole genomes, an interesting phenomenon is observed [Mrázek, 2009]. While the domains *Bacteria* and *Archaea* are clearly segregated based on the 16s rRNA distances, those domains are almost indistinguishable using the genome signature distances. Clearly, the signals obtained from the rest of the genome introduce significant noise resulting in a low signal-to-noise ratio. Although the signal to distinguish *Archaea* and *Bacteria* is present in 16s rRNA sequences, this signal is corrupted with noise.

Generally, the genome signatures are sufficient to infer correct taxonomic topologies. But with more quantitative criteria, they are weakly (yet significantly) correlated with the current reference distances (e.g., 16s rRNA distances in prokaryotes). This might be because of minimal usage of prior knowledge and averaging features with maximum uncertainity. On the other hand, molecular evolution is not homogeneously governed. As some sites evolve more rapidly than others, some patterns are conserved and some have higher flexibility. For instance, oligonucleotide content evolves at different rates. The most conserved 15 octanucleotides in *Bacteria* were reported to be homopurines and homoprymidines in similar rank [Davenport and Tümmler, 2010]. Conservation of certain homopurines and homoprymidines was claimed to be associated with the physical genome structure. On the other hand, the frequency of occurrence of some oligomers varied considerably even in close lineages. This leads us to the claim that variations in conserved oligomers correspond to ancient events, and variations in rapidly evolving oligomers correspond to relatively recent events. Therefore, assigning equal weights to all compositional elements results in the lack of molecular clocks in genome signature distance measurements. As a future work of genome-signature-based phylogeny construction, it is possible to design sequence distance measurements with trained molecular clocks. This can be performed with a careful optimization procedure taking a reference method (e.g., 16s rRNA sequence distances) into consideration. Consequently, utilizing prior knowledge using this scheme, greater phylogenetic accuracy could become attainable.

CHAPTER 5

Applications: Metagenomics

5.1 COMMUNITY ANALYSIS OF ENVIRONMENTAL SAMPLES

Microbic organisms are involved in numerous processes of life on Earth. Microorganisms are a source of nutrients, they cycle organic matter, and they form symbiotic relationships with life forms at every level of the tree of life. In aggregate, they make up a great proportion of the living population in the biosphere. While the microbial world dominates life on Earth and understanding this world is crucial for many areas, ranging from biological sciences to other fields such as medicine, agriculture, and the food industry, our current understanding of microbes is very limited. It is estimated that less than one percent of the microbial world has been explored [Pace, 1997, Rappe and Giovannoni, 2003]. This is primarily due to the technical limitations on isolation and culturing of microbes in nature. Only a small percentage of microbes can be cultured and studied by microbiologists. Thus, the current knowledge of microbiology is biased in favor of the small proportion of culturable species.

Since the sequencing of the first bacterial genome in 1995 [Fleischmann et al., 1995], genomes of more than 1,000 microbial species have been sequenced and annotated. This number is much less than the known minority of microbial diversity. Since it is not possible to isolate the majority of existing microbes, the current paradigm is not sufficient for extensively exploring the tree of life. It limits genomic analysis to the small percentage of the existing species which are culturable. The newborn science of **metagenomics**, often acknowledged as a paradigm shift in microbiology [National Research Council of the National Academies, 2007], has the potential to overcome the limitations on microorganism annotation. Metagenomics enables the genomic study of environmental samples; and thus it deals with the unknown majority of microbes for which isolation of single genomes is not possible [Achtman and Wagner, 2008, Hugenholtz, 2002]. A principal goal of metagenomics is to sample microbiomes and recover genetic material without isolating individual organisms.

Recovering genetic material *en masse* provides great opportunities for various areas of research. *In situ* sampling enables recovery of genetic material from various environments, such as ocean [DeLong, 2005, DeLong et al., 2006], soil [Daniel, 2005], hot springs [Barns et al., 1994], hydrothermal vents [Huber et al., 2002], polar ice caps [Christner et al., 2003], and hypersaline environments [Bellnoch, 2002]. This new type of complex data gathered from the environment directly requires a novel analysis approach as it introduces new research challenges. However, even the early techniques involving conventional genome analysis have revealed valuable insights. Exploring the taxonomic and metabolic diversity at the ecosystem level is one of the practical achievements of metagenomics. Analysis of environmental samples can also lead to advances in biotechnology [Lorenz and Eck, 2005, Schmeisser et al., 2007], the study of human physiology [Turnbaugh et al., 2007], and genetic archeology of extinct species [Noonan et al., 2006, Green et al., 2006]. Discovery of novel genes encoding biocatalysts and drugs, as well as the discovery of other biomolecules, can be counted as achievements of the early era of metagenomics [McHardy and Rigoutsos, 2007, Andersson et al., 2008, Tringe et al., 2005]. Eventually, advances in metagenomics should help to extend the tree of life [Doolittle, 1999] while enriching sequence libraries. Furthermore, the study would expand analysis from genomic to metagenomic: interactions within communities could be studied extensively using samples from various habitats.

5.2   SAMPLING AND SEQUENCING ENVIRONMENTAL SAMPLES

In order to gather genetic material from an environmental sample, the first step is to sample organisms from the environment. The goal of metagenome sampling is to obtain a sufficient number of chromosomes from each species existing in the microbial community. Population sizes of different operational taxonomic units (OTUs) in a mixture might be diverse, resulting in the underrepresentation of low populated species. This imposes a requirement on the number of chromosomes that should be gathered from the environment in order to achieve a complete representation of the community.

Rarefaction curves, which plot the number of OTUs gathered versus the number of individuals sampled, are used to determine the quality of sampling [Wooley et al., 2010]. As the slope of a rarefaction curve converges to zero, a complete representation of the microbial population is obtained. Ideally, many individuals can be sampled to guarantee a complete sampling of the metagenome. However, due to the increased cost and restricted budgets of metagenomics projects, an ideal sampling might not always be possible.

Following sampling, environmental samples are filtered. This is a physical procedure, and organisms in a sample are eliminated according to their physical size. The goal in many microbiology projects is to eliminate small viroids and large protists to obtain the sampled bacterial population in the corresponding habitat. There are other metagenome projects that are targeted to viromes [Willner et al., 2009]; and in this case, viral organisms are subject to the filtering process.

Whole shotgun sequencing of the recovered organisms is the next step required to obtain the genetic information. The product of sequencing depends on factors such as the sampling size, and the sequencing technology employed.

Depending on the diversity of the microbial community, an environmental sample can include from a few dominant species to thousands of species at the same level of dominance [Tyson et al., 2004, Garcia Martin et al., 2006, Strous et al., 2006]. Examples of low diversity metagenomes include the gutless worm symbiont community [Woyke et al., 2006], for which long contigs in the range of 100 kbp - 1 Mbp were assembled, and acid mine drainage biofilms [Tyson et al., 2004], in which complete genome assemblies of the dominant species were obtained. However, in diverse communities, only very short contigs are achievable. In termite hindgut microbiomes [Warnecke et al., 2007], soil, and whale fall (deep ocean) [Tringe et al., 2005] samples, contig assemblies do not exceed 10 kbp in length.

This missing data problem stems mostly from the sequencing constraints and project budgets. For the popular Sanger sequencing method [Sanger et al., 1975, 1977], metagenomic projects usually result in a total of 100 Mbp [Joint Genome Institute, 2000]. For a community of microbes with different abundance ratios, this amount of data will only cover relatively abundant sequences while the rest of the population will remain with insufficient coverage roughly proportional to their relative abundance in the population. A worse scenario exists for high diversity communities: none of the organisms will have enough coverage for the assembly of long contigs. This results in missing portions of the genomes in the sample and short sequences which are generally insufficient for analysis of genes and phylogenetic diversity [Wooley et al., 2010, Wommack et al., 2008].

The Lander-Waterman equation [Lander and Waterman, 1988] suggests that generation of longer total sequenced data will proportionally increase the average coverage per base. Given a properly sampled environmental sample, this would mean sufficient coverage to assemble organisms with lower abundance is possible in theory with production of massive amounts of sequencing output. With the introduction of high-throughput sequencing technologies, lower cost per base and faster sequencing is now possible [Mardis, 2008, Metzker, 2010, Fox, 2009]. Second generation sequencing technologies are replacing high-cost, labor-intensive Sanger sequencing. The Life Sciences 454-GS FLX Titanium 454 pyrosequencer [FLX, 2000] can produce 400 Mbp in a single run while the Illumina GAIIx [Illumina, 2000] can produce 15-20 Gbp per run, the $SOLiD^{TM}$ (Sequencing by Oligo Ligation and Detection) platform [SOLiD, 2000] can yield 20 Gbp per run and the single-molecule sequencing platform, $HelicosHeliScope^{TM}tSMS$ [Helicos, 2000] is capable of producing >1Gbp/hour. The feasibility of producing greater amounts of metagenome data has accelerated the area of metagenomics. It was reported that in the last five years, second generation sequencing has generated a greater amount of sequenced DNA than Sanger sequencing has generated in the last three decades [Metzker, 2010].

5.3   EXPLORATION OF BIODIVERSITY IN A METAGENOME

For ideal phylogenetic and functional genomics analysis, complete genomes are needed. In practice, fragmented genomes in long contigs can also be very informative for various levels of analysis. However, this is only currently achievable for dominant species in low diversity populations. This lack of ability to obtain sufficiently long contigs from individual genomes in a microbial mixture has forced researchers to approach the metagenome as a "bag of genes" and conduct the analysis on a gene level. Phylogenetic diversity is usually explored by characterizing OTUs using PCR amplification of marker genes, such as 16S rRNA genes [DeSantis et al., 2006], or using non-rRNA genes [Case et al., 2007, von Mering et al., 2007].

Multiple housekeeping genes are used in multilocus sequence typing (MLST) for exploring phylogenetic diversity [Mahenthiralingam et al., 2006]. Unfortunately, approaches which estimate phylogenetic diversity using marker genes are known to have several problems [Suzuki and Giovannoni, 1996, von Wintzingerode et al., 1997]. Recently, a core set of marker genes was determined to be used in phylotyping. AMPHORA [Wu and Eisen, 2008] and MLTreeMap [von Mering et al., 2010] analyze these marker genes to infer the phylogenetic information of a given environmental sample. While these programs supply information about the biodiversity of a sample, they only associate those genome fragments that carry a marker gene with possible OTUs. This means that the great majority of sequencing reads remain unassociated with any taxa. Table 5.1 shows the percentage of the DNA sequences in a metagenome mixture which are assigned to taxa using phylotyping methods. The reason for this poor assignment is that only a small part of a genome contains the marker genes, and this dramatically reduces the probability of occurrence a marker gene for a given random genome fragment. In fact phylotyping approaches suggest answers for the question "what groups are in the mix?" rather than the question of "Which fragment belongs to which one of the groups in the mixture?"

Table 5.1: Table 5.1: The percentage of fragments assigned to taxa in a metagenome using marker gene-based phylotyping methods.

Method	16s RNA	MLST	AMPHORA	MLTreeMap
Assignment (%)	< 1	< 1	1.3	1.89

5.4   METAGENOME ASSEMBLY

Metagenome assembly is the process of obtaining long contigs or drafts of complete genomes from sequence reads. The sequenced metagenomes include fragment reads of multiple genomes from various organisms existing in the environment. An ideal scenario for the assembly of genomes populating the metagenome would be assembling each genome in parallel fashion after a taxonomical classification phase [Salzberg and Yorke, 2005, Johnson and Slatkin, 2006]. Realizing such an approach is currently an open research problem.

The contemporary approach to metagenome analysis is to employ taxonomic grouping after attempts to assemble the metagenome treating it as a single species read set. There are several problems with this approach. Taxonomic classification operates successfully with the sequences having a length in the long contig range. On the other hand, attempting to assemble long contigs without taxonomic grouping, or binning, leads to poor assemblies. This is the conundrum of metagenomics data analysis: a good assembly of genomes in an environmental sample requires phylogenetic classification, while good phylogenetic classification requires assembled contigs of sufficient length and thus containing significant information for characterization. To date, no comprehensive metagenome assembler has been reported; and conventional genome assemblers are facing difficulties with data consisting of a mixture of several genomes, which eventually affects the performance of taxonomic classification. Currently, single genome assembly programs, such as Forge, Phrap [PHRAP, 2000], TIGR, CAP3 [Huang and Madan, 1999], Arachne [Batzoglou et al., 2002, Jaffe et al., 2003], JAZZ

[Aparicio et al., 2002], the Celera Assembler [Myers et al., 2000], and EULER [Pevzner et al., 2001, Chaisson and Pevzner, 2008], are also employed for metagenome assembly [Pop, 2009]. These programs are specifically designed for Sanger sequencing and the assembly of isolated genomes. Modifications to these algorithms adapting them to perform on the greater number of shorter reads yielded by new generation sequencing are also available with the programs such as SSAKE [Warren et al., 2007], VCAKE [Jeck et al., 2007], SHARCGS [Dohmet al., 2007], Velvet [Zerbino and Birney, 2008], and Allpaths.

5.5   METAGENOME BINNING

One of the computational tasks in metagenome analysis called binning, involves categorizing sequenced data into OTUs for further analysis. Binning is a difficult problem when the information required for differentiation has to be obtained from short DNA reads. A number of approaches have been proposed for computational binning of metagenome data, and some of them are currently employed in real-life metagenome analysis.

It is possible to categorize the binning approaches in three main classes: similarity search methods, supervised compositional methods, and unsupervised methods. While the first category involves molecular database searches for previously explored homogenous sequences, the latter two use the notion of genome signatures to bin the DNA sequences to taxa.

5.5.1   SIMILARITY SEARCH-BASED BINNING METHODS

Probably the most widespread method of binning is using homology searches for a given unknown genomic fragment. As mentioned earlier, using a few marker genes is insufficient to label a great majority of metagenomical fragments. Employing a larger set of molecular sequences has been shown to facilitate of metagenome binning. Here, by larger sets of molecular sequences we refer to comprehensive sets of protein sequences and assemblies of whole genomes, or large contigs from known organisms. Corresponding molecular data gathered from various projects have been deposited in public databases. Consequently, the task of searching for matches between unknown metagenome samples and known sequences reduces to homology searches in molecular databases.

An example of employing homology search using known protein domains is the algorithm CARMA [Krause et al., 2008]. CARMA assigns sequences to taxonomical origins by trying to match them to known protein families contained in Pfam domains. Profile Markov models are used to search the aligned Pfam domains for possible homologies. Although this class of methods is frequently used for phylotyping, they can be employed for binning since they comprehensively compare protein domains and attempt to classify any given genome fragment. While computationally expensive, CARMA has been shown to be accurate even for short sequences in the current pyrosequencing read length range (80-400 bp). However, the accuracy drops dramatically when phylogenetically close sequences are missing from the search databases. Running CARMA on a comprehensive dataset gathered from a large spectrum of known genomes resulted in inaccurate classifications

[Krause et al., 2008] (6% sensitivity when using 100 bp sequences for identification at the genus level).

Another similarity-based method is MEGAN [Huson et al., 2007, 2009], which uses the scores of similarity searches to assign the DNA fragments to taxa using a lowest common ancestor algorithm. Usually nucleotide BLAST [Altschul, 1997] is employed for the similarity search task. Therefore, a common binning strategy using MEGAN appears to be a local alignment search using available DNA sequences of known organisms. MEGAN is reported to be successful when the organisms forming the metagenome have close relatives in the search databases. However, in a recent study [Dinsdale, 2008], only 12% of the data obtained from microbial communities in coral atolls got significant BLAST hits. SOrt-ITEMS [Monzoorul et al., 2009] is a recent example of similarity search methods employing BLAST as an ontology search strategy. In addition to similarity search scores, the search parameters are also considered in the taxonomy assignment algorithm.

Similarity search methods are very powerful when homologous sequences exist in search databases, because significant hits with local alignments are expected to have high ratios of true positives. However, in practice, homology searches are unable to identify sequences from a large proportion of the microbial population. The reason for this is the small number of sequenced biological molecules compared to the vast number of species in metagenome samples. Poor identification results have been reported with real-life metagenome data as a matter of course.

5.5.2   SUPERVISED COMPOSITIONAL BINNING METHODS

Supervised compositional binning methods approach the problem of binning from a general perspective of modeling. According to this scheme, genome fragments are represented as compact mathematical models which represent the species-specific characteristics of genomes. Sequenced genomes in public databases are also represented by their models. The homology search task of sequence similarity-based methods is replaced with model comparison. The model-based approach provides several advantages: first, the computational burden is reduced when compared to similarity based methods, and second the models provide a more general representation. The reduction in computational burden is a crucial practical issue in metagenomics analysis, since large amounts of data have to be processed, which might result in infeasibility problems. Similarity based methods require sequence alignment runs over voluminous databases, whereas, supervised compositional binning methods generally compare relatively small structures. Moreover, the representation of sequences by structures that emphasize the specific features provides a concise framework. Introduction of a more general scheme has been observed to be more accurate for a number of binning scenarios [Brady and Salzberg, 2009].

Genome signatures, being species specific and pervasive, are a plausible candidate for DNA sequence modeling to be employed in supervised binning methods. While the specific character helps in distinguishing fragments from different genome sources, the pervasiveness enables the use of the signature with short fragments usually seen in metagenomes.

The naive-Bayesian-Classifier-based method proposed by Sandberg et al. and a Markov chain method by Dalevi et al. are early examples of this approach. The algorithm PhyloPythia [McHardy et al., 2007] consists of various support vector machine (SVM) classifiers. Relative frequency profiles of short oligonucleotides (5-mers for clade levels of genus to class, and 6-mers for the clade levels of phylum and domain) were used as feature vectors. Relative oligonucleotide frequency vectors were generated for various fragment lengths, and SVMs were trained using different fragment lengths. Satisfactory sensitivity and specificity results are reported for the sequence lengths > 1-3kbp. However, a sharp cut-off in the accuracy is observed for fragments less than 1 kbp in length. Another recent taxonomic classification method, TACOA [Diaz et al., 2009], proposes a k-nearest-neighbor-classification-based algorithm. In this method, genomic sequences are represented by over- underabundance profiles of oligonucleotides called genomic feature vectors (GFV). GFVs are identical to zero'th order Markov models. Training GFVs over known genomes, the best score calculated from the closest k trained neighbors to a test GFV determines the taxonomic assignment of an unknown test query. For sequence lengths under 1 kbp, 4-mers are used to build GFVs. For longer sequences, the frequencies of 5-mers are observed to perform the best. TACOA has been shown to correctly classify fragments larger than 800 bp with an average sensitivity between 76% at the rank of superkingdom and 39% at the rank of genus. Its performance is comparable to PhyloPythia in that range. As the distance metric, Euclidean distances are used and fed into radial basis functions in PhyloPythia, whereas inner products are used in TACOA.

Phymm [Brady and Salzberg, 2009] was developed for the classification of short read lengths of metagenomics data. It is based on a Bayesian decision machine which detects the taxonomic source of a read with its maximum a posteriori probability calculated over variable order Markov models. Complete genomes of known taxa are used for training Markov models. Oligonucleotide lengths of 1-mer to 8-mer are used in training the models. The use of Phymm significantly increases accuracy compared to CARMA and PhyloPhytia.

5.5.3 UNSUPERVISED METHODS

The previous two classes of binning methods require the prior knowledge of sequence information for known taxonomic units. When the majority of species embodying a metagenome is included in model or sequence databases, the binning performance is satisfactory. When unidentified and nonsequenced genomes exist in the mixture, the taxonomic classification becomes impossible. Given contemporary limited knowledge of microbial sequences, this is not an unexpected scenario. Furthermore, discovery of new microbes is conceptually very limited with the similarity-search-based and supervised methods. Since supervised and similarity-based binning methods label the metagenome with known species, the exploration is confined to the small portion of the known microworld, or its close relatives.

For discovery of novel microbial species, unsupervised categorization of metagenomes is needed. The requirement for unsupervised binning is the ability to distinguish fragments of different sources without the aid of trained models. That is to say, accurate clustering of metagenome samples has to be achieved. Employing genome signatures within an autonomous framework of categorization appears to be an appropriate approach to unsupervised binning.

5.5.3.1 Unsupervised Binning Using Self-Organizing Maps

Early examples of unsupervised binning, made use of autonomous neural network structures called self organizing maps (SOM) [Kohonen, 1982, Kohonen et al., 1996]. SOMs group similar structures using batch learning methods which minimize the mean square classification error. SOMs are useful for the visualization of high-dimensional data; they project the complex relation of data onto a simple two-dimensional map.

The possibility of clustering metagenome samples using genome signatures was extensively investigated in [Dick et al., 2009]. It was previously reported that genomes sharing the same environment are similar in composition [Foerstner et al., 2005, Willenbrock et al., 2006, Raes et al., 2007, Paul et al., 2008]. As organisms in a metagenome share the same environment, this could result in a problem of disappearance of species-specific features of genome signatures in metagenomes. However, a case study performed on an acid mine metagenome in which the organisms share extremely acidic conditions has shown that the genome signatures are not obscured. Using SOMs as the clustering scheme and tetranucleotide frequencies of 5 kbp fragments, a clear clustering of metagenome samples was observed. Moreover, specificity was observed for fragments as short as 500 bp, and clusters form around the length of 1400 bp.

[Abe et al., 2005] reported a clear separation of species with 1 kb and 10 kb fragments from 65 prokaryotes and 6 eukaryotes using 2,3,4-mer oligonucleotide frequencies. Later, they supported their results using clinical data from uncultured microbes. Comparing the clusters with the known genomes, they concluded that 79% of the Sargasso Sea metagenome consists of unknown species.

Different architectures of SOMs further improved the binning results of this class of unsupervised methods. Using growing self-organizing maps, hyperbolic SOMs in unsupervised [Chan et al., 2008a, Martin et al., 2008] and semisupervised settings [Chan et al., 2008b], accuracy values comparable with supervised binning were achieved.

5.5.3.2 Binning Methods Considering Community Abundance

The diversity of populations, leading to under- and overabundance of species in a microbial community affect the clustering characteristics of metagenome binning. If a taxon has an abundant number of individuals, the variance of signatures within a taxon might be large, compared to an inter-taxa variance of low abundance sequences. Different approaches which take into account the population abundance have been implemented in a number of binning programs.

Compostbin [Chatterji et al., 2008] uses data reduction with weighted principal component analysis. The 4^6 dimensional feature space of hexanucleotide frequencies calculated for each fragment is reduced down to three dimensions of the largest principal components. The weighting scheme first estimates the coverage of sequences by fast approximate sequence alignment [Kent, 2002] and the inverse of the coverage assigned to each fragment as the weighting factor. The final distance graph is partitioned using bisection by normalized cuts. Performing the bisections iteratively results in binning clusters.

LikelyBin [Kislyuk, 2009] estimates the genome signatures in the form of Markov models and incorporates them with *a priori* probability of each fragment which is proportional to the abundance value of the related organism in the metagenome mix. A Markov chain Monte Carlo setting estimates the corresponding probabilities (i.e. genome signatures and population abundance) simultaneously. Consequently, the *a posteriori* probabilities of a fragment for each model indicates the cluster to which the fragment belongs. AbundanceBin [Wu and Ye , 2010] is an expectation-maximization algorithm which uses the Lander-Waterman model. Oligomer frequency estimates are used for maximization of the *a posteriori* probability of an oligonucleotide coming from a certain species. Once the algorithm converges, the estimated values are used for sequence binning. Tetra [Teeling, 2004] is one of the earliest tools used to group the fragments in a metagenome. It uses relative proportions of tetranucleotides with respect to the database samples in DNA contigs and calculates the correlations of pairs as a measure of similarity. In [Yang et al., 2010], only some of the oligonucleotides, which are believed to carry the phylogeny information, are used for metagenome binning. An approach filtering oligonucleotides which occur with similar frequency between different DNA fragments, as well as the ones with extremely different occurrence statistics, improves the binning results. SCIMM [Kelley and Salzberg, 2010] is the unsupervised version of the program Phymm. Interpolated Markov models are trained for metagenome fragments and clustered using an expectation maximization algorithm which maximizes the likelihood functions. SCIMM was compared with LikelyBin and CompostBin implementations, and improvement in clustering results was reported.

CHAPTER 6

Applications: Horizontal DNA Transfer Detection

We have reviewed several applications of computational genomic signatures involving the evolutionary distances between genomes and detection of the taxonomical membership of genome fragments. These applications exploit the species specificity and pervasiveness of genomic signatures. As reported on several occasions, these properties are not ideally observed in real genomes. Many genome signatures define mathematical characterizations of genomes based on the assumption that a DNA sequence is generated by a stationary random process. However, genomes are not homogenous; and they are composed of several regions with different characteristics.

An important reason for atypical regions and heterogeneity in genomes is a phenomenon called horizontal transfer. A horizontal transfer event leads to the acquisition of DNA material by one species from another. The transfer of genetic material between different species may not only create an evolutionary leap in the genotype of a species, but it may significantly shape the phenotype of that species. Detection of these horizontal transfers is very important for a better understanding of evolutionary processes, as well as for conducting medical and industrial research.

Applications of computational genomic signatures to determine horizontal transfers have great importance, because they constitute plausible approaches with simplicity and accuracy. Similar to other genome-signature-based applications, the homogeneity of a given genome is assumed. However, this time it is also assumed that genomes harbor alien regions that are horizontally transferred from other species. Species specificity and pervasiveness suggest that the alien regions have distinct composition since they are not native to the host genome, and it is possible to detect alien regions based on these composition differences. Therefore, the application of genome signatures for the detection of horizontal transfers includes anomaly detection approaches. In this chapter, we first review horizontal transfer events; then we discuss the corresponding genome signature applications.

6.1 HORIZONTAL GENE TRANSFER

The horizontal transfer of genes between two different species was reported as early as 1944, when the transformation of nonvirulent to virulent *Streptococcus pneumoniae* [Avery et al., 1944] was observed. This phenomenon was considered to be a rare virulent event causing pathogenies, and the genome regions carrying horizontally transferred DNA were named pathogenicity islands by Hacker et al. to describe a horizontally transferred region of the uropathogenic *Escherichia coli* genome [Hacker et al., 1990]. Later, it was discovered that the horizontal transfer of genes is not limited to virulence factors; and it occurs more generally for nonpathogenic microbes, including *Archaea* genomes [Garcia-Vallve et al., 2000]. The nonvirulent genomic islands are associated with adaptive metabolic activity genes [Hacker and Kaper, 2000], antibiotic resistance genes, or secretion system genes [Hentschel and Hacker, 2001]. In this context, the horizontally transferred genes contribute to the adaptation of an organism to its environment, thus they are positively selected. In fact, the habitat-related adaptation tendencies provide more horizontal transfer opportunities to nonpathogenic species compared to parasitic microbes [Garcia-Vallve et al., 2000].

The transfer of genetic material to a host genome can happen in several ways. Conjugation of DNA can be a reason for horizontal transfers [Grohmann et al., 2003]. The DNA transfer in this fashion requires the physical contact of cells with the help of secretory systems. Phage transduction is another possible cause of DNA transfers between host cells and temperate or lysogenic phages [Zinder and Lederberg, 1952]. This requires the integration of phage genome to the host genome after entering the host cell. In some cases the integration of prophages to the host genome, instead of lysing the cell, is observed to be a fitness-increasing event [Brussow et al., 2004]. Another mechanism, transformation of DNA [Dubnau, 1999], is a direct uptake of DNA fragments from the environment and integration with genome or plasmids in the cell. In addition to the horizontal transfer between different cellular organisms is the intracellular DNA material exchange between genomes and plasmids or genomes and organelle genomes. The integration of transferred DNA fragments to the host genome is mainly executed using the mechanisms of transposition, homologous recombination, or other integron mechanisms [Zaneveld et al., 2008].

Recent studies show that horizontal transfer events are apparent and significant in the evolutionary history of prokaryotes [Koonin et al., 2001]. The discovery of frequent homologies between unrelated species [Maynard-Smith and Smith, 1998, Lecointre et al., 1998] showed that genetic transfers are not rare events taking place in specific conditions. In fact, horizontally transferred regions form a large percentage of prokaryote genomes. These percentages might change from 0.5% in the endocellular symbiont *Buchnera* sp. APS genome to up to 25% in the euryarchaeal *Methanosarcina acetivorans* genome, with an average of around 15% [Nakamura et al., 2004]. These estimates are considered to be very conservative. This is because it is not possible to determine the ancient horizontally transferred regions which have become indistinguishable from native DNA using current detection tools. Moreover, during the course of evolution, several horizontal transfer events do not leave footprints in the contemporary genomes since the related regions are lost. For example, it is estimated that *E. coli* and *Salmonella enterica* lineages have gained and lost 3 Mbp of sequences in the last 100 million years [Lawrence and Ochman, 1997]. Even with these conservative estimates, horizontal gene transfers seem to significantly affect evolutionary history. With the most conservative estimates, it would take around 10,000 years with mutational evolution mechanisms to obtain the same novelty obtained in around one year with horizontal transfers [Simonson et al., 2005].

The frequent occurrence of horizontal transfer events, introduction of important differences in the genetic material composition, and significant genome percentage coverage of genomic islands indicate the importance of horizontal transfers in the history of evolution. Recent findings show that horizontal gene transfer is one of the driving forces of evolution. The diversity introduced by the vertical evolution factors, such as point mutations, insertions, and deletions, introduce small effects for adaptation to the environment, which are than carried by the horizontal genetic transfers [Lawrence, 1999]. Horizontal gene transfers are accepted to be one of the main reasons for the flux in genomes, along with gene loss, duplication, gene fusion, fission, and genesis [Snel et al., 2002]. The heterogeneity of genomes due to horizontally transferred genes results in inconsistencies in phylogeny constructions. When gene-based phylogeny construction is considered, use of different genes would alter the trees [Simonson et al., 2005]. The reason is that horizontally transferred genes appear to be the representatives of taxa which are different from the genome under investigation. This results in one challenge in describing taxonomic relationships using individual genes of organisms. On the other hand, especially in prokaryotic phylogeny construction, use of 16sRNA sequences is the most common method for phylogeny construction. Ribosomal RNA sequences are well conserved and good representatives of the evolutionary divergence processes experienced. However, they indicate only the vertical evolution driven mostly by point mutations. Horizontal transfers, which increase the fitness of an organism with acute metabolic gains, are discarded with the use of marker genes. Moreover, it is argued that horizontal gene transfers erase the pyhlogenetic history recorded by the DNA in prokaryotic genomes [Doolittle, 1999]. Because of the importance of horizontal transfer events it has been proposed by several researchers that a reevaluation of taxonomic relationships be considered [Gevers et al., 2005, Bapteste, 2005, Embley and Martin, 2006, Ochman et al., 2005, Ge et al., 2005]. There has even been some questions as to whether if it is even possible to construct the tree of life as proposed by Darwin [Darwin, 1887].

It is possible to categorize genes according to their tendency to be transferred to host genomes. Jain, Rivera, and Lake investigated the gene orthology relationships using the operational genes from a set of prokaryotes [Jain et al., 1999]. Starting from the ancient horizontal transfers to the recent ones, operational genes were observed to be transferred much more frequently than the informational genes. Here the genes responsible for the core structural activities, such as transcription and translation, are called the informational genes. Operational genes are the ones that are responsible for metabolic events, such as amino acid biosynthesis. This leads to the complexity hypothesis which claims that informational genes are not transferred frequently, since they are a part of large-scale structural systems forming complex networks. On the other hand, operational genes code for small functional elements which can be mobilized and potentially increase the fitness of an organism. It was also claimed that operational genes related to the cell surface, DNA binding, and pathogenicity are transferred more frequently [Nakamura et al., 2004].

6.1.1   HORIZONTAL GENE TRANSFER IN PROKARYOTES

Horizontal transfers in prokaryotic and eukaryotic organisms exhibit different characteristics. The possibility and frequency of horizontal transfer events in prokaryotes are higher [de la Cruz and Davies, 2000]. The single-cell, plastic, and adaptive organisms of the microbial world utilize horizontal transfers relatively more frequently than most higher order Eukaryotic organisms. It is possible to observe these transfers in each stage of the evolutionary history of prokaryotes. The transfers, that had taken place in early ancestors appear to be ancient horizontal transfers. However, it is also possible to observe relatively recent transfers which distinguish different strains of the same species. For example, different strains of *H. pylori* have a distinct phenotype because of horizontally transferred genomic islands. The related acquisitions are recent ones, and they bring out different metabolic capabilities for strains of the same species. Also, the close species *E.coli* and *Salmonella enterica* are distinguished from each other by horizontal transfers [Lawrence and Ochman, 1997].

Among prokaryotes, nonpathogenic *Bacteria* and *Archaea* contain a larger percentage of horizontally transferred islands in their genomes compared to parasitic bacteria. It is thought that habitat features enable a greater opportunity to exchange genetic materials.

The horizontal transfers in prokaryotes can crucially affect the lifestyle of an organism. As an example, hyperthermophilic bacteria *Aquifex aeolicus* and *Thermotoga maritima*, contain a massive amount of horizontally transferred genes from *Archaea*. These species share the same lifestyle with *Archaea*, possibly as a result of the metabolic capabilities they have acquired from the Archaea [Aravind et al., 1998, Nelson et al., 1999].

6.1.2   HORIZONTAL GENE TRANSFER IN EUKARYOTES

Horizontal gene transfer events in eukaryotic genomes happens rarely compared to prokaryotic organisms. It is thought that most transfers happen by the insertion of viruses and retroviruses into eukaryote genomes [Van Blerkom, 2003, Hedges and Batzer, 2005], via exchange of genetic materials involving symbionts or parasites [Dunning Hotopp et al., 2007], by the transfer from organelle genomes in intracellular fashion [Doolittle et al., 2003, Bock and Timmis, 2008], or by hybridization [Burke and Arnold, 2001].

The horizontal gene transfer phenomenon in eukaryotes has not been studied as intensively as the transfer events in prokaryotes and is thus not well understood. Nevertheless, the importance of horizontal transfer events in the evolution of eukaryotes has been considered. Horizontal gene transfer among eukaryotic species in different clade levels and transfers within the same genus [Silva and Kidwell, 2000] and between different genera [Won and Renner, 2003] have been reported. Moreover, the transfer between different kingdoms has also been observed in eukaryotic genomes [Veronico et al., 2001, Intieri and Buatti, 2001, Mallet et al., 2010].

6.1.2.1   Genesis of Eukaryotes: a Fusion Hypothesis

There are several hypotheses for the appearance of nuclear genomes of eukaryotes. An analysis considering the horizontal gene transfers from prokaryotes attempts to construct phylogeny trees using three Bacteria, three Archaea, and two eukaryotes [Simonson et al., 2005]. The trees generated using the gene sets attain poor resolution. However, when the individual trees are aligned together allowing cyclic permutations, perfect matches can be achieved. As a result, instead of attempting to construct a tree of life, a graph allowing multiple parents results in a consistent scenario. According to this result, a fusion of either a proteobacterium or a photosynthetic clade that included the Cyanobacteria and the Proteobacteria, with an archaeal eocyte. Homology detections also show that the informational genes of eukaryotic nuclear DNA are rooted in Archaea, whereas the operational genes of these genomes are rooted in Bacteria. This fusion hypothesis is an example indicating how important horizontal transfers can be in the evolutionary history of species.

6.2   HORIZONTAL GENE TRANSFER DETECTION

Detection of horizontally transferred genomic islands and the genes they harbor carry scientific and practical information for life sciences. Firstly, as a major force of evolution, it is crucial to determine the horizontal transfers between different taxa. A better understanding of genetic material exchange will illuminate the complex relationships between species, and it will help build the tree of life. Second, the virulence factors resulting in pathologies need to be revealed genetically for medical studies. This requires careful detection of pathogenicity islands placed in the host genomes of pathogenic organisms. Moreover, decoding the mechanisms changing the phenotype of organisms and helping the metabolic adaptations to different environmental conditions is an application of horizontal transfer detection. Determination of these mobile metabolic factors is also potentially important for industrial applications, microbiology, and ecological studies.

6.2.1   COMPARATIVE METHODS

A conventional approach to detecting horizontally transferred genes in a genome is to conduct homology searches using a database of biological sequences. In this approach, any detected homology from another taxonomic group proposes a horizontal transfer in the previous history of the given organism. That is to say, in order to predict a gene or island which is transferred to the genome, similar sequences have to be previously known.

Early approaches to finding horizontally transferred genes in prokaryotic genomes included BLAST searches with manual curation [Koonin et al., 2001]. Homology search methods also became popular in detecting the pathogenicity islands [Yoon et al., 2005]. BLAST searches with protein databases were also used to detect the prophage sequences lying in genomes [Srividhya et al., 2006]. Some comparative methods use phylogeny trees built using gene sequences and detect transfers by the inconsistencies in the trees [Santos and Ochman, 2004]. Most comparative methods require manual curation with expert knowledge. Recently, automated horizontal transfer detection methods have been proposed. MobilomeFINDER [Ou et al., 2007], MOSAIC [Chiapello et al., 2005], and IslandPick [Langille et al., 2008] are some popular methods that use genome alignments to detect the genomic islands. Darkhorse [Podell and Gaasterland, 2007] automates the phylogenetic tree approach by weighting individual trees according to the prior knowledge of taxonomic relatedness of species.

Although comparative methods can be very accurate when close homologous sequences are available in the databases, they bring many limitations. Firstly, a very comprehensive or-thologous analysis of proteins is required for each gene under investigation [Smith et al., 1992,

Poptsova and Gogarten, 2007]. This means, even for recent horizontal transfers, multiple strains of a species have to be sequenced and annotated. Even with the increasing rate of sequencing and genome annotation, many sequences do not have ortologuous counterparts within the annotated minority of species. Another disadvantage of comparative methods is that it is difficult to conclude if a sequence inconsistency between two close relatives is due to gene loss or horizontal transfer. Finally, comparative methods require a massive amount of sequence alignments, which results in significant computational burden.

6.2.2   METHODS BASED ON GENOME SIGNATURES

Since alien islands acquired from other taxa possibly have different compositions than the native sequences of the host genome, performing compositional analysis is another approach to horizontal transfer detection. Genome signatures extract compositional signals native to a genome efficiently. Therefore, use of genome signatures is a plausible candidate as a tool for compositional analysis.

Besides its simplicity and accuracy, horizontal transfer detection using genome signatures has other advantages. No comparison with other genomes and ortologuous sequences are required to carry out a genome-signature-based horizontal transfer analysis. This provides opportunity for *ab initio* detection of horizontal transfers. In addition, unlike most comparative methods, analysis with genome signatures is not limited to gene sequences. Noncoding regions of genomes can be included in the analysis. Since horizontal transfers happen with the exchange of genomic islands, which harbor a significant amount of noncoding fragments, additional information in these regions can be utilized.

Some examples of horizontal gene detection categorized based on the type of genomic signatures can be summarized as follows.

6.2.2.1   GC Content and Codon Usage

Since compositional feature analyses based on GC content and synonymous codon usage can be viewed as variations of genome signatures, we can include them in the class of genome-signature-based horizontal transfer detection approaches. Detection based on GC content anomaly or genes having atypical synonymous codon usage have been reported to be capable of detecting certain horizontal transfers [Zhang and Zhang, 2004, Garcia-Vallve et al., 1999, Merkl, 2004]. Methods using GC content deviation employ either gene by gene analysis [Garcia-Vallve et al., 1999] or use sliding windows [Garcia-Vallve et al., 1999, Karlin, 2001, Tu and Ding, 2003]. Some GC content methods consider the GC content in different codon positions separately [Lawrence and Ochman, 1997, 1998]. Methods employing synonymous codon usage use individual genes by definition using different distance metrics, such as χ^2 distance [Lawrence and Ochman, 1998], δ^* distances [Karlin, 2001, Tu and Ding, 2003], Kullback-Leibler divergence [Hayes and Borodovsky, 1998], and Mahalanobis distances [Garcia-Vallve et al., 1999].

6.2.2.2 Dinucleotide Abundance Ratios

Dinucleotide abundance ratios which are one of the best known genome signatures, and their variations are frequently employed for horizontal transfer detection. [Karlin, 2001] used the dinucleotide abundance ratios with the δ^* distance to determine the pathogenicity islands in bacteria. Later, this approach was used for determining genomic islands [Tu and Ding, 2003] and was employed in the genomic island analysis programs IslandPath [Hsiao et al., 2003] and *compare_islands* [van Passel et al., 2005].

6.2.2.3 Higher Order Oligonucleotide Signatures

The use of longer oligonucleotides (i.e., ≥ 4) have been implemented in several horizontal transfer detection methods. This includes different oligonucleotide lengths with different distance metrics. Chaos Game Representation signatures defined with tetranucleotides using Euclidean metrics have been used for horizontally transferred region detection in [Dufraigne et al., 2005], and a similar method was used for determining the genomic flux of *Helicobacter pylori* strains [Saunders et al., 2005] and detecting horizontal transfers in fungi [Mallet et al., 2010]. Composition-vector-based atypical genomic region browser SeqWord [Ganesan et al., 2008] uses tetranucleotide frequencies for detection. Frequencies of longer oligonucleotides have been proposed for use in a few programs. W_n [Tsirigos and Rigoutsos, 2005a] is an algorithm using 8-mer and 9-mer frequencies for each genome and determining the horizontally transferred genes to be the genes most distant from the centroid of the signature determined by the entire set of the genes. This approach is further improved by employing support vector machines for anomaly detection [Tsirigos and Rigoutsos, 2005b]. Alienhunter [Vernikos and Parkhill, 2006] is a genomic island detection algorithm which uses interpolated variable-order Markov models derived from 8-mers and Kullback-Leibler divergence as the distance metric. A fragment in a moving window is compared to the model of the entire genome in this scheme.

Use of longer oligonucleotide frequencies is also integrated with other genomic features, such as insertion point location, size of the genomic region, gene density of the considered fragment, existence of repeats in the fragment, and existence of integrase-like protein domains in the fragment, phage-related protein domain, or RNA sequences. GIDetector uses a decision tree to decide if a given fragment is a horizontally transferred genomic island [Dongsheng et al., 2010]. The same features are combined in a Relevance Vector Machine (RVM) in [Vernikos and Parkhill, 2008].

6.3   PERFORMANCE AND LIMITATIONS OF GENOME SIGNATURES FOR HORIZONTAL TRANSFER DETECTION

Prediction of horizontally transferred parts of genomes is difficult. Furthermore, no gold standards exist to evaluate the performance of horizontal transfer detection methods [Lawrence and Ochman, 2002, Ragan et al., 2006]. Different methods provide distinct results. This is attributed to the fact that the property and the features emphasized by each signature scheme corresponds to a different aspect of horizontal transfers, and this might lead to varying detection success for different types of horizontal transfers [Ragan, 2001]. This is another difficulty for quantifying the accuracy of different methods. However, it is possible to compare the performance of different methods by using a standard dataset and measuring the sensitivity and specificity characteristics of existing methods. For this purpose, a set of standard artificial genome is used in [Becq et al., 2010]; and existing methods were compared using these datasets. In Figure 6.1, the sensitivity and specificity characteristic curves for these methods using the standard genomes are shown.

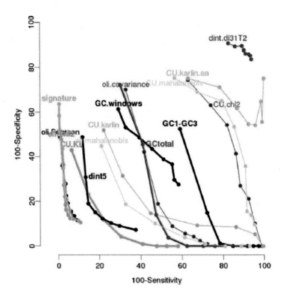

Figure 6.1: Sensitivity-specificity operating characteristic curves for different genome-signature-based horizontal transfer detection methods. A curve placed closer to the origin is more accurate than the others.(Figure taken from [Becq et al., 2010].)

It can be seen from the characteristic curves that for the most part longer oligonucleotide-based methods are closer to the origin. Codon usage and CG content methods tend to have worse performance. In addition to the structure of the characterization, it can be seen that the metric used to measure the differences of signatures also have an effect on the impact the overall accuracy. For example, when codon usage is used with Kullback-Leibler divergence, its performance becomes comparable to methods using tetranucleotide frequencies. When the total performance is considered, tetranucleotide frequencies with the KL divergence metric are observed to be the most successful.

These observations are in accordance with our previous observation about genome signatures. Use of longer oligonucleotides enables the gathering of extra information from longer context, which includes longer-term dependencies. On the other hand, careful selection of distance metrics results in more effective use this information. Here, an example of information theoretic distance metrics appears to be the most successful one.

When a genome fragment is received by a host genome, the composition of the host genome and the alien fragment are different depending on the evolutionary distance of the donor and the host. However, in later stages of evolution, the transferred fragment is exposed to the intracellular direct mutational pressures, which are major factors driving the genome signature of the host. Therefore, the signature of the transferred region converges to the host signature. This process is generally called amelioration [Lawrence and Ochman, 1997]. Ancient transfers, which have been significantly ameliorated, are almost indistinguishable from the host composition, thus they are difficult to detect [Nakamura et al., 2004]. Similar to the ancient received fragments, recent transfers with evolutionarilly close donors are also hard to detect because of the close compositions of the host and the donor genomes. Conversely, an atypical region is not necessarily a recent genome transfer. Some ancient transferred genes have higher resistance to amelioration and their compositions change slowly. That is to say no standard molecular clock exists for amelioration and the atypical behavior of genomic island composition is not donor independent [Linz et al., 2000]. Because of this complexity, it is hard to draw conclusions about the detected horizontal transfers. However, we can say that what determines the performance of horizontal transfer detection is the absolute compositional differences rather than the chronology of transfers or the relationship of the donors. Therefore, not surprisingly, genome signatures with greater species specificity and pervasiveness appear to be more successful at detecting transfers because they have higher distinguishing ability both between close and distant relatives. Alienhunter was reported to be the most sensitive method for the detection of ancient transfers [Langille et al., 2008]. The reason behind its sensitivity is possibly the use of a strong signature with longer oligonucleotide frequency functions (8-mers) and an information theoretic distance measure (KL divergence).

6.3.1   COMPOSITIONAL SIMILARITY OF HOST AND DONOR GENOMES

The relative closeness of donor genome composition appears to be one of the major contributors to the horizontal transfer detection accuracy. A key observation on this issue is that most species, especially prokaryotes, tend to exchange genetic material with organisms from the same habitat. That is to say the donors and hosts are likely to have similar genome signatures, since organisms sharing the same environment tend to have similar genome compositions [Foerstner et al., 2005]. It is possible to define some environmental parameters and observe the correlation of the oligonucleotide frequency-based signatures with these parameters. An observable set of parameters could include the optimal growth temperature (measured in celsius), respiratory behavior (categorized as aerobic, anaerobic, and facultative), and habitat (categorized as aquatic, host-associated, multiple, specialized, and terrestrial). In Figure 6.2, the classification accuracy (habitat and oxygen requirement) and the correlation coefficient (optimal growth) temperature is shown for mono- to pentanucleotide frequency vectors. Those values are calculated using 10-fold cross-validation employing support vector classifiers and regressors. The values at zero oligomer length correspond to random classification (or zero correlation). An increasing trend in the accuracy with increasing length of the oligonucleotide is clear. This is an indication that the compositional effects of the environmental factors effect bases close together and for statistics gathered from a large context of a nucleotide in the sequence are more informative about the organism's environment.

This shaping effect of the environment on the genome compositions and, consequently, on the genome signatures, implies that there is a higher higher probability that the horizontal transfer of species sharing a similar environment also means the horizontal transfer of DNA with close signatures. We can interpret this phenomenon in a number of different ways. It might be proposed that genome fragments with similar signatures tend to adapt to each other easier. That is why there might be positive selection for horizontal transfers with donor DNA having similar composition. However, there are several examples of horizontal transfer among different kingdoms with significantly dissimilar genomic composition. A neutral view is that this tendency stems from geographical and functional constraints. Since organisms sharing the same environment have a higher probability of physical contact, they have more opportunity to exchange genetic material. Moreover, similar environmental parameters requires similar metabolic strategies for adaptation. This means that the fitness increasing genes that an organism needs have a higher probability of existing in the same environment. All in all, the tendency to transfer genetic material from the same environment is a potential difficulty for horizontal transfer detection, since it might result in small signature distances between DNA sequences.

6.3.2   OTHER CHALLENGES LIMITING THE PERFORMANCE OF GENOME-SIGNATURE-BASED HORIZONTAL TRANSFER DETECTION

Amelioration and the transfer of fragments with similar compositions seems to be the major difficulties of genome- signature-based transfer detection methods. There are other challenges stemming from atypical composition within a genome which are not due to horizontal transfers. For example, pseudogenes having different evolutionary pressures than the operational genes within a genome might have different composition. Similarly, genes under mutational pressures different from the general trends of the genome can also result in having atypical compositions [Garcia-Vallve et al., 2000]. An example of this situation is the composition of ribosomal RNA genes. RNA signatures tend to be different than the genome average, and they form a different cluster in the signature space [Dufraigne et al., 2005, Tsirigos and Rigoutsos, 2005a]. Different mutational pressures on the genes placed on the leading and the lagging strands leads to atypical signatures [Lafay et al., 1999, Baran and Ko, 2008]. The synonymous codon usage patterns are known to alter with gene expression levels. For this reason, highly expressed genes also have a high probability of having atypical composition, which might result in false positives in horizontal detection.

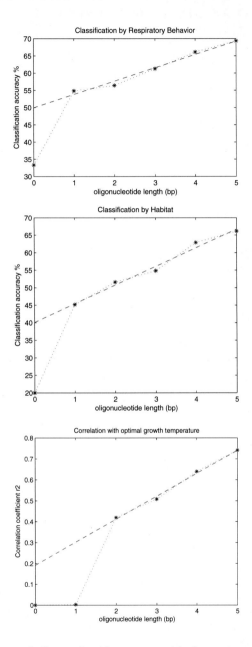

Figure 6.2: The correlations of oligonucleotide content with the environmental parameters. It can be seen that similar environmental factors results in similar genomic compositions. The increase in the oligonucleotide length results in better correlations.

6.3.3 AMELIORATION AND DEVIATION FROM GENERAL GENOMIC SIGNATURE TRENDS

The main rationale behind the detection of horizontal transfers using genome signatures is to determine the alien regions by making use of their compositional atypicality with respect to the host genome. This approach considers the current image of the genome composition, thus it does not use any features from the evolutionary history of the horizontal transfer. However, studies on the amelioration of horizontally transferred genes imply that there is more information contained in the composition of these regions [Lawrence and Ochman, 1997, Baran and Ko, 2008].

[Lawrence and Ochman, 1997] developed a back-amelioration algorithm to determine how long ago a horizontal transfer took place. According to their scheme, an amelioration rate for each of the three codon positions should be estimated. Because of the lower pressure on the third codon position, it was found that the GC content of the third codon ameliorates much faster than the GC content of the first and the second codon positions. According to Muto and Osawa (MO) relationships, the GC content of three codon positions are highly correlated with the overall GC content of a genome. Moreover, the regression relationships are almost constant through phyla. That means, given the average GC content of a genome, it is possible to estimate the GC content of all three codon positions. However, this is only true for the genomes in equilibrium. A horizontally transferred gene is close to the equilibrium if it is recently transferred. The reason for that is the composition of the alien fragment will be very similar to the donor composition, which is in equilibrium. On the other hand, an ancient transfer which is extensively ameliorated can be considered to be in equilibrium, because its composition has been driven to a state similar to the host signature. Lawrance and Ochman observed that the transferred regions in the process of amelioration are not in equilibrium and they diverge from the general trends of Muto and Osawa relationships. An unequal rate of amelioration for three codon positions disables a linear transition. Therefore, the context-dependent GC distributions are observed to be outliers with respect to the general trends of genomes. The back-amelioration algorithm uses the estimated amelioration rates and simulates the reverse process of amelioration iteratively, until the Muto and Osawa relationships are met. This corresponds to the time of horizontal transfer. The estimated amelioration rates help to predict the horizontal transfer time. The introduction times of *cob* and *lac* operons of *Salmonella enterica* have been estimated by the back-amelioration algorithm. The estimates are in accordance with the findings in the literature.

The results of Lawrence and Ochman imply that the horizontally transferred fragments under the amelioration process carry distinct signals which can be exploited for detection. Genome signatures have strict distribution in the space they are represented. The first few principal components of the oligonucleotide frequency vectors, sampled from the currently sequenced genomes, contain most of the total variance. This is an implication that the genomes in equilibrium follow a certain trend which can be represented by a few components. Moreover, the GC content of genomes is strongly correlated with the first principal component of the oligonucleotide frequency vectors [Deschevanne et al., 1999]. The universal trends observed for synonymous codon usage are also supportive in this context. The synonymous codon usage of a genome can be represented with a few parameters including GC content. Muto and Osawa relationships are a projection of these trends for synonymous codon usage, since the GC content for each position of codons can be derived from codon usage. More generally, these relationships are projections of trends in genome signatures. Therefore, it is possible that the deviation from Muto and Osawa relationships is a consequence of the deviation from the general trends of signatures during the process of amelioration [Muto and Osawa, 1987]. Therefore, algorithms emphasizing the deviation from equilibrium signatures can be developed for detecting sequences under the amelioration process. This is a potential use of genome signatures, extracting additional information from genome composition, which can lead to improvement in horizontal gene transfer detection.

Bibliography

C. H. Cannon, C. S. Kua, E. K. Lobenhofer, P. Hurban, Capturing genomic signatures of DNA sequence variation using a standard anonymous microarray platform, *Nucleic Acids Research*, 34. Art. No. e121, 2006. DOI: 10.1093/nar/gkl478 1

K. J. Livak, S. J. Flood, J. Marmaro, W. Giusti, K. Deetz, Oligonucleotides with fluorescent dyes at opposite ends provide a quenched probe system useful for detecting PCR product and nucleic acid hybridization, *PCR Methods Appl*, 4: 357-362, 1995. 1

A. M. Phillippy, J. A. Mason, K. Ayanbule, D. D. Sommer, E. Taviani, A. Huq, R. R. Colwell, I. T. Knight, S. L. Salzberg, Comprehensive DNA signature discovery and validation, *PLoS Comput Biol*, 3(5):e98, 2007. DOI: 10.1371/journal.pcbi.0030098 1

P. D. Hebert, M. Y. Stoeckle, T. S. Zemlak, C. M. Francis, Identification of birds through DNA barcodes, *PLoS Biol*, 2:e312, 2004. DOI: 10.1371/journal.pbio.0020312 1, 52

R. D. Ward, T. S. Zemlak, B. H. Innes, P. R. Last, P. D. Hebert, DNA barcoding Australia's fish species, *Philos Trans R Soc Lond B Biol Sci*, 360:1847-1857, 2005. DOI: 10.1098/rstb.2005.1716 1, 52

M. Vences, M. Thomas, A. van der Meijden, Y. Chiari, D. R. Vieites, Comparative performance of the 16S rRNA gene in DNA barcoding of amphibians, *Frontiers in Zoology*, 2: article 5, 2005a. DOI: 10.1186/1742-9994-2-5 1, 52

S. E. Miller, DNA barcoding and the renaissance of taxonomy, *Proc Natl Acad Sci USA*, 104:4775-4776, 2007. DOI: 10.1073/pnas.0700466104 1

K. H. Chu, C. P. Li, J. Qi, Ribosomal RNA as molecular barcodes: A simple correlation analysis without sequence alignment, *Bioinformatics*, 22:1690-1701, 2006. DOI: 10.1093/bioinformatics/btl146 2, 40

K. H. Chu, C. P. Li, Rapid DNA barcoding analysis of large datasets using the composition vector method, *BMC Bioinformatics*, 10(Suppl 14):S8, 2009. DOI: 10.1186/1471-2105-10-S14-S8 2, 40, 59

A. Nakabachi, A. Yamashita, H. Toh, H. Ishikawa, H. E. Dunbar, et al., The 160-kilobase genome of the bacterial endosymbiont Carsonella, *Science*, 314:267, 2006. DOI: 10.1126/science.1134196 2

K. Y. Lee, R. Wahl, E. Barbu, Contenu en bases puriques et pyrimidiques des acides desoxyribonu-cleiques des bacteries, *Ann. Inst. Pasteur.*, 91, 212-224, 1956. 2

H. Ochman, J. G. Lawrence, Phylogenetics and the amelioration of bacterial genomes, *Pp. 2627-2637 in F. C. Neidhardt, R. Curtiss III, J. L. Ingraham, E. C. C. Lin, K. B. Low, B.*, 1996. DOI: 10.1007/PL00006158 2

W. Saenger, *Principles of Nucleic Acid Structure*, Springer-Verlag, New York, 1984. 2

L. J. Marnett, Oxyradicals and DNA damage, *Carcinogenesis*, 21(3):361-370, 2000. DOI: 10.1093/carcin/21.3.361 3

E. Freese, On the evolution of base composition of DNA, *J. Theor. Biol.*, 3, 82-101, 1962. DOI: 10.1016/S0022-5193(62)80005-8 3

N. Sueoka, On the genetic basis of variation and heterogeneity of DNA base composition, *Proc. Natl. Acad. Sci. USA.*, 48:582-592, 1962. DOI: 10.1073/pnas.48.4.582 3

R. Cavicchioli, Cold-adapted archaea, *Proc. Natl. Acad. Sci. USA.*, 4:331-343, 2006. DOI: 10.1038/nrmicro1390 4

D. A. Hickey, G. A. C. Singer, Genomic and proteomic adaptations to growth at high temperature, *Genome Biol.*, 5(10):117, 2004 DOI: 10.1186/gb-2004-5-10-117 4

J. G. Bragg, C. L. Hyder, Nitrogen versus carbon use in prokaryotic genomes and proteomes, *Proc Biol Sci.*, 271(Suppl 5):S374-S377, 2004. DOI: 10.1098/rsbl.2004.0193 4

J. G. Bragg et al., Variation among species in proteomic sulphur content is related to environmental conditions, *Proc. Biol. Sci.*, 273, 1293-1300, 2006. DOI: 10.1098/rspb.2005.3441 4

B. S. Berlett, E. R. Stadtman, Protein oxidation in aging, disease, and oxidative stress, *J. Biol. Chem.*, 272:20313-20316, 1997. DOI: 10.1074/jbc.272.33.20313 4

R. Grantham, Workings of the genetic code, *Trends Biochem. Sci.*, 5, 327-331, 1980. DOI: 10.1016/0968-0004(80)90143-7 4

S. Osawa, T. H. Jukes, K. Watanabe, A. Muto, Recent evidence for evolution of the genetic code, *Microbiol. Rev.*, 56, 229-264, 1992. 4

M. Gouy, C. Gautier, Codon usage in bacteria: correlation with gene expressivity, *Nucleic Acids Res*, 10:7055-7074, 1982. DOI: 10.1093/nar/10.22.7055 4

T. Xie, D. Ding, X. Tao, D. Dafu, The relationship between synonymous codon usage and protein structure [published erratum appears in FEBS Lett 1998 Oct 16;437(1-2):164], *FEBS Lett*, 434:93-96, 1998. DOI: 10.1016/S0014-5793(98)00955-7 4

A. Carbone, A. Zinovyev, and F. Kepes, Codon Adaptation Index as a measure of dominating codon bias, *Bioinformatics*, 19:2005-2015, 2003. DOI: 10.1093/bioinformatics/btg272 4

G. D'Onofrio, D. Mouchiroud, B. Aïssani, C. Gautier, G. Bernardi, Correlations between the compositional properties of human genes, codon usage, and amino acid composition of proteins, *Journal of molecular evolution*, 32(6):504-510, 1991. DOI: 10.1007/BF02102652 4

J. R. Lobry, Influence of genomic G+C content on average amino-acid composition of proteins from 59 bacterial species, *Gene*, 205:309-316, 1997. DOI: 10.1016/S0378-1119(97)00403-4 4

P. G. Foster, L. S. Jermiin, D. A. Hickey, Nucleotide composition bias affects amino acid content in proteins coded by animal mitochondria, *J Mol Evol.*, Mar;44(3):282-8, 1997. DOI: 10.1007/PL00006145 4

N. Sueoka, Compositional correlation between deoxyribonucleic acid and protein, *Cold Spring Harb. Symp. Quant. Biol*, 26:35-43, 1961. DOI: 10.1101/SQB.1961.026.01.009 4

A. A. Adzhubei, I. A. Adzhubei, I. A. Krasheninnikov, S. Neidle, Nonrandom usage of "degenerate" codons is related to protein three-dimensional structure, *FEBS Lett*, 399:78-82, 1996. DOI: 10.1016/S0014-5793(96)01287-2 4

S. K. Gupta, S. Majumdar, T. K. Bhattacharya, T. C. Ghosh, Studies on the relationships between the synonymous codon usage and protein secondary structural units, *Biochem Biophys Res Commun*, 269:692-696, 2000. DOI: 10.1006/bbrc.2000.2351 4

S. L. Chen, W. Lee., A. K. Hottes, L. Shapiro, H. H. McAdams, Codon usage between genomes is constrained by genome-wide mutational processes, *Proc. Natl. Acad. Sci.*, 101: 3480-3485, 2004. DOI: 10.1073/pnas.0307827100 4, 12

R. Knight, S. Freeland, L. Landweber, A simple model based on mutation and selection explains trends in codon and amino-acid usage and GC composition within and across genomes, *Genome Biology*, 2(4): research0010.1-0010.13, 2001. DOI: 10.1186/gb-2001-2-4-research0010 4, 13

G. A. Palidwor, T. J. Perkins, X. Xia, A general model of codon bias due to GC mutational bias, *Genome Biology*, Oct 27;5(10):e13431, 2010. DOI: 10.1371/journal.pone.0013431 4, 13

M. Zama, Codon usage and secondary structure of mRNA, *Nucleic Acids Symp Ser*, 22:93-94, 1990. 4

A. Carbone, F. Kepes, A. Zinovyev, Codon bias signatures, organization of microorganisms in codon space, and lifestyle, *Mol Biol Evol*. 22:547-561, 2005. DOI: 10.1093/molbev/msi040 4, 13

J. Josse, A.D. Kaiser, and A. Kornberg, Enzymatic synthesis of deoxyribonucleic acid: VIII. frequencies of nearest neighbor base sequences in deoxyribonucleic acid, *Journal of Biological Chemistry*, 236:864-875, 1961. 5

M. N. Swartz, T. A. Trautner, and A. Kornberg, Enzymatic synthesis of deoxyribonucleic acid: XI. further studies on nearest neighbor base sequences in deoxyribonucleic acid, *Journal of Biological Chemistry*, 237:1961-1967, 1962. 5

J. H. Subak-Sharpe, Base doublet frequency patterns in the nucleic acid and evolution of viruses, *British Medical Bulletin*, 23:161-168, 1967. 5

G.J. Russell and J. H. Subak-Sharpe, Similarity of the general designs of protochordates and invertebrates, *Nature*, 266:533-536, 1977. DOI: 10.1038/266533a0 5

J. M. Morrison, H. M. Keir, J. H. Subak-Sharpe, L. V. Crawford, Nearest neighbour base sequence analysis of the deoxyribonucleic acids of a further three mammalian viruses: Simian virus 20, human papilloma virus and adenovirus type 2, *Journal of General Virology*, 1:101-108, 1967. DOI: 10.1099/0022-1317-1-1-101 5

C. Burge, A. M. Campbell, S. Karlin, Over- and under-representation of short oligonucleotides in DNA sequences, *Proceedings of the National Academy of Sciences*, 89:1358-1362, February 1992. DOI: 10.1073/pnas.89.4.1358 5, 6, 39

S. Karlin, L. R. Cardon, Computational DNA Sequence Analysis, *Annual Review of Microbiology*, 48:619-654, 1994a. DOI: 10.1146/annurev.mi.48.100194.003155 5, 39

S. Karlin and I. Ladunga, Comparison of eukaryotic genomic sequences, *Proceedings of the National Academy of Sciences, USA*, 91:12832-12836, December 1994b. DOI: 10.1073/pnas.91.26.12832 5, 6, 39, 53

S. Karlin, E. S. Mocarski, and G. A. Schachtel, Molecular Evolution of Herpesviruses - Genomic and Protein Sequence Comparison, *Journal of Virology*, 68:1886-1902, March 1994c. 5, 39

S. Karlin, I. Ladunga, and B. E. Blaisdell, Heterogeneity of genomes: Measure and values, *Proceedings of the National Academy of Sciences, USA*, 91:12387-12841, 1994d. DOI: 10.1073/pnas.91.26.12837 5, 39

S. Karlin and C. Burge, Dinucleotide relative abundance extremes: A genomic signature, *Trends in Genetics*, 11:283-290, July 1995. DOI: 10.1016/S0168-9525(00)89076-9 5, 6, 39

S. Karlin, J. Mrazek, and A. M. Campbell, Compositional Biases of Bacterial Genomes and Evolutionary Implications, *Journal of Bacteriology*, 179:3899-3913, June 1997a. 5, 39

S. Karlin and J. Mrazek, Compositional differences within and between eukaryotic genomes, *Proc. Natl. Acad. Sci. USA*, Vol. 94, pp. 10227-10232, September 1997b. DOI: 10.1073/pnas.94.19.10227 5

S. Karlin, A. M. Campbell, and J. Mrazek, Comparitive DNA Analysis across Diverse Genomes, *Annual Review of Genetics*, 32:185-225, December 1998. DOI: 10.1146/annurev.genet.32.1.185 5, 39

F. L. Bai, Y. Z. Liu, T. M. Wang, A representation of DNA primary sequences by random walk, *Math Biosci*, Sep;209(1):282-91, 2006. DOI: 10.1016/j.mbs.2006.06.004 8

M. A. Gates, Simple DNA sequence representations, *Nature*, 316:219, 1985. DOI: 10.1038/316219a0 8

M. Leong, S. Morgenthalar, Random walk and gap plots of DNA sequences, *Comput. Appl. Biosci.*, 21:503, 1995. DOI: 10.1093/bioinformatics/11.5.503 8

M. Kowalczuk, P. Mackiewicz, D. Mackiewicz, A. Nowicka, M. Dudkiewicz, M. R. Dudek, S. Cebrat, High correlation between the turnover of nucleotides under mutational pressure and the DNA composition, *BMC evolutionary biology*, 17:1-13, 2001. DOI: 10.1186/1471-2148-1-13 8

H. J. Jeffrey, Chaos game representation of gene structure, *Nucleic Acids Research*, 18:2163-2170, 1990. DOI: 10.1093/nar/18.8.2163 8

R. L. Devaney, *An Introduction to Chaotic Dynamical Systems*, Addison Wesley, Redwood City, California, 1989. 8

P. J. Deschevanne, A. Giron, J. Vilain, A. Vaury, and B. Fertil, Genomic signature: Characterization and classification of species assessed by chaos game representation of sequences, *Molecular Biology and Evolution*, 16:1391-1399, 1999. 8, 9, 10, 39, 82

P. Deschavanne, A. Giron, J. Vilain, C. Dufraigne, B. Fertil, Genomic signature is preserved in short DNA fragments, *BIBE 2000 IEEE International Symposium on bio-informatics and biomedical engineering. Washington, D.C., USA*, 2000. DOI: 10.1109/BIBE.2000.889603 9, 39

F. Tekaia and E. Yeramian, Evolution of proteomes: fundamental signatures and global trends in amino acid compositions, *BMC Genomics* 7:307, 2006. DOI: 10.1186/1471-2164-7-307 13

Y. Wang, K. Hill, S. Singh, and L. Kari, The spectrum of genomic signatures: From dinucleotides to chaos game representation, *Gene*, 346:173-185, 2005. DOI: 10.1016/j.gene.2004.10.021 9

R. W. Jernigan and R. H. Baran, Pervasive properties of the genomic signature, *BMC Genomics*, 3, 23, 2002. DOI: 10.1186/1471-2164-3-23 15

C. K. Peng, S. V. Buldyrev, A. L. Goldberger, S. Havlin, F. Sciortino, M. Simons, H. E. Stanley, Long-range correlations in nucleotide sequences, *Nature*, Mar 12;356(6365):168-70, 1992. DOI: 10.1038/356168a0 16

W. Li and K. Kaneko, Long-range correlation and partial $1/f^{\alpha}$ spectrum in a noncoding DNA sequence, *Europhys. Lett.*, 17, 655-660, 1992. DOI: 10.1209/0295-5075/17/7/014 16

R. F. Voss, Evolution of long-range fractal correlations and 1/f noise in DNA base sequences, *Phys. Rev. Lett.*, 68, 3805-3808, 1992. DOI: 10.1103/PhysRevLett.68.3805 16

S. Karlin and V. Brendel, Patchiness and correlations in DNA sequences, *Science,* 259, 677-680, 1993. DOI: 10.1126/science.8430316 16

S. V. Buldyrev, A. L. Goldberger, S. Havlin, R. N. Mantegna, M. E. Matsa, C. -K. Peng, M. Simons, and H. E. Stanley, Long-range correlation properties of coding and noncoding DNA sequences: GenBank analysis, *Phys. Rev.,* E 51, 5084-5091, 1995. DOI: 10.1103/PhysRevE.51.5084 16

D. Holste, I. Grosse, S. Beirer, P. Schieg, H. and Herzel, Repeats and correlations in human DNA sequences, *Phys. Rev.,* E 67, 061913, 2003. DOI: 10.1103/PhysRevE.67.061913 16

H. Herzel, W. Ebeling, A. Schmitt, Entropies of biosequences: the role of repeats, *Phys. Rev.,* E 50, 5061- 5071, 1994. DOI: 10.1103/PhysRevE.50.5061 17

D. Holste, I. Grosse, S. Buldyrev, H. Stanley, H. Herzel, Optimization of coding potentials using positional dependence of nucleotide frequencies, *J. Theor. Biol.,* 206, 525-537, 2000. DOI: 10.1006/jtbi.2000.2144 17

I. Grosse, P. Bernaola-Galva´n, P. Carpena, R. Roma´n-Rolda´n, J. Oliver, H. Stanley, Analysis of symbolic sequences using the Jensen-Shannon divergence, *Phys. Rev.,* E 65, 041905, 2002. DOI: 10.1103/PhysRevE.65.041905 17

M. Berryman, A. Allison, D. Abbott, Mutual information for examining correlations in DNA, *Fluctuation Noise Letters,* 4, 237-246, 2004. DOI: 10.1142/S0219477504001847 17, 23

P. Jacobs and P. Lewis, Discrete time series generated by mixtures III: autoregressive processes (DAR(p)), *Tech. Rep. NPS55-78-022, Naval Postgraduate School,* Monterey, California, 1978. 17

P. Jacobs and P. Lewis, Stationary discrete autoregressive-moving average time series generated by mixtures, *J. Time Ser. Anal.,* 4, 19- 36, 1983. DOI: 10.1111/j.1467-9892.1983.tb00354.x 17, 18

M. Dehnert, W. E. Helm, M.-T. Hutt, A discrete autoregressive process as a model for short-range correlations in DNA sequences, *Physica A,* 327, 535-553, 2003. DOI: 10.1016/S0378-4371(03)00399-6 17, 39

M. Dehnert, W. E. Helm, M.-T. Hutt, Information theory reveals large scale synchronisation of statistical correlations in eukaryote genomes, *Gene,* 345:81-90, 2005a. DOI: 10.1016/j.gene.2004.11.026 18, 39

M. Dehnert, R. Plaumann, W. E. Helm, M. T. Hutt, Genome phylogeny based on short-range correlations in DNA sequences, *J Comput Biol,* Jun;12(5):545-53, 2005b. DOI: 10.1089/cmb.2005.12.545 23, 39, 59

M. Dehnert, W. E. Helm, M.-T. Hutt, Informational structure of two closely related eukaryote genomes, *Physical Review E,* 74:021913-1-021913-9, 2006. DOI: 10.1103/PhysRevE.74.021913 18, 39

C. Shannon, A Mathematical Theory of Communication, *Bell Syst Tech J,* 27:379-423, 623-65, 1948. DOI: 10.1145/584091.584093 20, 30, 32, 45

B. Korber, R. Farber, D. Wolpert, A. Lapedes, Covariation of Mutations in the V3 Loop of Human Immunodeficiency Virus Type I Envelope Protein: An Information Theoretic Analysis, *Proc Natl Acad Sci,* 90:7176-7180, 1993. DOI: 10.1073/pnas.90.15.7176 20

R. Roman-Roldan, P. Bernaolo-Galvan, J. Oliver, Theory to DNA Sequence Analysis: A Review, *Pattern Recognition,* 29(7):1187-1194, 1996. DOI: 10.1016/0031-3203(95)00145-X 20

B. Giraud, A. Lapedes, L. Liu, Analysis of Correlations Between Sites in Models of Protein Sequences, *Phys Rev E,* 58(5):6312-6322, 1998. DOI: 10.1103/PhysRevE.58.6312 20

H. Herzel, I. Grosse, Correlations in DNA Sequences: The Role of Protein Coding Segments, *Phys Rev E,* 55:800-810, 1997. DOI: 10.1103/PhysRevE.55.800 20

I. Hofacker, M. Fekete, P. Stadler, Secondary structure prediction for aligned RNA sequences, *Journal of Molecular Biology,* 319:1059-1066, 2002. DOI: 10.1016/S0022-2836(02)00308-X 20

S. Lindgreen, P. Gardner, A. Krogh, Measuring covariation in RNA alignments: physical realism improves information measure, *Bioinformatics,* 22:2988-2995, 2006. DOI: 10.1093/bioinformatics/btl514 20

M. Bauer, S. Schuster, K. Sayood, The average mutual information profile as a genomic signature, *BMC Bioinf.,* 9:48, 2008. DOI: 10.1186/1471-2105-9-48 21

H. Herzel, O. Weiss, E. N. Trifonov, 10-11 bp periodicities in complete genomes reflect protein structure and DNA folding, *Bioinformatics,* Mar;15(3):187-93, 1999. DOI: 10.1093/bioinformatics/15.3.187 23

E. N. Trifonov and J. L. Sussman, The pitch of chromatin DNA is reflected in its nucleotide sequence, *PNAS,* 77, 3816-3820, 1980. DOI: 10.1073/pnas.77.7.3816 23

S. L. Salzberg, Microbial gene identification using interpolated Markov models, *Nucleic Acids Res,* 26:544-548, 1998. DOI: 10.1093/nar/26.2.544 23

A. L. Delcher, K. A. Bratke, E. C. Powers, S. L. Salzberg, Identifying bacterial genes and endosymbiont DNA with Glimmer, *Bioinformatics,* 23, 673-679, 2007. DOI: 10.1093/bioinformatics/btm009 23

D. Dalevi, D. Dubhashi, and M. Hermansson, Bayesian classifiers for detecting HGT using fixed and variable order markov models of genomic signatures, *Bioinformatics,* 22:517-522, March 2006. DOI: 10.1093/bioinformatics/btk029 24, 42

I. Saeed, S. K. Halgamuge, The oligonucleotide frequency derived error gradient and its application to the binning of metagenome fragments, *BMC Genomics*, Dec 3;10 Suppl 3:S10, 2009. DOI: 10.1186/1471-2164-10-S3-S10 27, 39

S. Vinga, and J. S. Almeida, Renyi continuous entropy of DNA sequences, *J. of Theor. Biol.*, 231, 37788, 2004. DOI: 10.1016/j.jtbi.2004.06.030 30

S. Vinga, and J. S. Almeida, Local Renyi entropic profiles of DNA sequences, *BMC Bioinformatics*, 8, 393, 2007. DOI: 10.1186/1471-2105-8-393 30

J. Ziv, and A. Lempel, A universal algorithm for sequential data compression, *IEEE Transactions on Information Theory*, 23, 337-343, 1977. DOI: 10.1109/TIT.1977.1055714 30, 58

J. Ziv, and A. Lempel, Compression of individual sequences via variable-rate coding, *IEEE Transactions on Information Theory*, 24, 530-536, 1978. DOI: 10.1109/TIT.1978.1055934 30, 46

K. Sayood, *Introduction to Data Compression*, Morgan Kauffman-Academic Press, San Francisco, 2005. 30

X. Chen, S. Kwong, and M. Li, A compression algorithm for DNA sequences and its applications in Genome comparison, *RECOMB 00: Proc. of the 4th Annual International Conference on Computational Molecular Biology*, 107-117, 2000. DOI: 10.1145/332306.332352 30

X. Chen, S. Kwong, and M. Li, Compression of DNA sequences, *Proc. IEEE Symp. on Data Compression*, 340-350, 1993. DOI: 10.1109/DCC.1993.253115 30

T. Matsumoto, K. Sadakane, and H. Imai, Biological sequence compression algorithms, *Genome Informatics*, 11, 43-52, 2000. 30

B. Behzadi, and F. L. Fessant, DNA compression challenge revisited: A dynamic programming approach, *CPM*, 190-200, 2005. DOI: 10.1007/11496656_17 30

M. D. Cao, T. I. Dix, L. Allison, and C. Mears, A simple statistical algorithm for biological sequence compression, *In Proc. of the IEEE Data Compression Conference (DCC), IEEE Computer Society*, 43-52, 2007. DOI: 10.1109/DCC.2007.7 30

G. Bejerano, and G. Yona, Variations on probabilistic suffix trees: statistical modeling and prediction of protein families, *Bioinformatics*, 17, 2343, 2001. DOI: 10.1093/bioinformatics/17.1.23 31

I. Ulitsky, D. Burstein, T. Tuller, and B. Chor, The average common substring approach to phylogenomic reconstruction, *J. of Computational Biology*, 13, 336-350, 2006. DOI: 10.1089/cmb.2006.13.336 31, 57

P. Grassberger, Estimating the information content of symbol sequences and efficient codes, *IEEE Trans. Inf. Theory*, 35, 669-675, 1989. DOI: 10.1109/18.30993 31

T. M. Cover and J. A. Thomas, Estimating the information content of symbol sequences and efficient codes, *Elements of Information Theory*, 2nd Edition Wiley Series in Telecommunications and Signal Processing, New York, 2006. DOI: 10.1109/18.30993 31

L. Gatlin, Triplet frequencies in DNA and the genetic program, *J. Theor. Biol.*, 5, 360-371, 1963. DOI: 10.1016/0022-5193(63)90083-3 31

L. Gatlin, The information content of DNA, *J. Theor. Biol.*, 10, 281-300, 1966. DOI: 10.1016/0022-5193(66)90127-5 31

L. Gatlin, The information content of DNA II, *J. Theor. Biol.*, 18, 181-194, 1968. DOI: 10.1016/0022-5193(68)90160-4 31

B. Schoelkopf and A. Smola, *Learning with Kernels*, MIT Press, Cambridge MA, 2002. 38

J. Bohlin, E. Skjerve, D. W. Ussery, Analysis of genomic signatures in prokaryotes using multinomial regression and hierarchical clustering, *BMC Genomics*, 10: 487, 2009a. DOI: 10.1186/1471-2164-10-487 40

J. Bohlin, E. Skjerve, Examination of genome homogeneity in prokaryotes using genomic signatures, *PLoS One*, 2;4(12):e8113, 2009b. DOI: 10.1371/journal.pone.0008113 40

J. Bohlin, E. Skjerve, D. W. Ussery, Investigations of oligonucleotide usage variance within and between prokaryotes, *PLoS Comput Biol.*, 18;4(4):e1000057, 2008. DOI: 10.1371/journal.pcbi.1000057 40

J. Mrázek, Phylogenetic signals in DNA composition: limitations and prospects, *Mol Biol Evol.*, May;26(5):1163-9, 2009. DOI: 10.1093/molbev/msp032 40, 60

H. Teeling, J. Waldmann, T. Lombardot, M. Bauer, F. O. Glockner, Tetra: a web-service and a stand-alone program for the analysis and comparison of tetranucleotide usage patterns in DNA sequences, *BMC Bioinformatics*, 5: 163, 2004. DOI: 10.1186/1471-2105-5-163 39, 40, 69

J. Qi, B. Wang, B. I. Hao, Whole proteome prokaryote phylogeny without sequence alignment: a K-string composition approach, *J Mol Evol.*, 58(1):1-11, 2004. DOI: 10.1007/s00239-003-2493-7 54

H. Wei, J. Qi, B. Hao, Prokaryote phylogeny based on ribosomal proteins and aminoacyl tRNA synthetases by using the compositional distance approach, *Sci China C Life Sci.*, 47(4):313-21, 2004. DOI: 10.1360/03yc0137

K. H. Chu, J. Qi, Z. G. Yu, V. Anh, Origin and phylogeny of chloroplasts revealed by a simple correlation analysis of complete genomes, *Mol Biol Evol.*, 21(1):200-6, 2004. DOI: 10.1093/molbev/msh002 55

T. Abe, H. Sugawara, S. Kanaya, T. Ikemura, A novel bioinformatics tool for phylogenetic classification of genomic sequence fragments derived from mixed genomes of uncultured environmental microbes, *Polar Biosci*, 20:103-112, 2006. 39

C.-K. K. Chan, A. L. Hsu, S. L. Tang, S. K. Halgamuge, Using Growing Self-Organising Maps to Improve the Binning Process in Environmental Whole-Genome Shotgun Sequencing, *Journal of Biomedicine and Biotechnology*, Article ID 513701:10, 2008a. DOI: 10.1155/2008/513701 39, 68

T. Kohonen, The self-organizing map, *Proc. IEEE*, 78, 1464-1480, 1990. DOI: 10.1109/5.58325

F. Zhou, V. Olman, Y. Xu, Barcodes for genomes and applications, *BMC Bioinformatics*, 9: 546, 2008. DOI: 10.1186/1471-2105-9-546 39

T. Abe, H. Sugawara, M. Kinouchi, S. Kanaya, T. Ikemura, Novel phylogenetic studies of genomic sequence fragments derived from uncultured microbe mixtures in environmental and clinical samples, *DNA Res*, 12:281-290, 2005. DOI: 10.1093/dnares/dsi015 39, 68

R. Sandberg, G. Winberg, C.-I. Brnden, A. Kaske, I. Ernberg, J. Coster, Capturing whole-genome characteristics in short sequences using a naive bayesian classifier, *Genome Research*, 11:1404-1409, 2001. DOI: 10.1101/gr.186401 41

J. Rissanen, and G. G. Langdon, Jr., Universal modeling and coding, *In IEEE Trans. Inform. Theory*, vol. IT-27, pp. 12-23, Jan. 1981. DOI: 10.1109/TIT.1981.1056282 45

C. Woese, O. Kandler, M. Wheelis, "Towards a natural system of organisms: proposal for the domains Archaea, Bacteria, and Eucarya.", *Proc Natl Acad Sci USA*, 87 (12): 4576-9, 1990. DOI: 10.1073/pnas.87.12.4576 49

C. Woese, O. Kandler, M. Wheelis, Updating prokaryotic taxonomy, *J Bacteriol*, 187:6255-6257, 2005. DOI: 10.1128/JB.187.18.6255-6257.2005 49

K. T. Konstantinidis, A. Ramette, J. M. Tiedje, The bacterial species definition in the genomic era, *Phil Trans Soc B*, 361:1929-1940, 2006. DOI: 10.1098/rstb.2006.1920 49

R. Rossello-Mora and R. Amann, The species concept for prokaryotes, *FEMS Microbiol Rev*, 25:39-67, 2001. DOI: 10.1111/j.1574-6976.2001.tb00571.x 49, 50

E. Stackebrandt, W. Frederiksen, G. M. Garrity, P. A. D. Grimont, P. Kampfer, et al., Report of the ad hoc committee for the re-evaluation of the species definition in bacteriology, *Int J Syst Evol Microbiol*, 52:1043-1047, 2002. DOI: 10.1099/ijs.0.02360-0 49, 51

J. Staley, The bacterial dilemma and the genomic-phylogenetic species concept, *Phil Trans Soc B*, 361:1899-1909, 2006. DOI: 10.1098/rstb.2006.1914 49

E. Chatton, *Titres et travaux scientiflques*, Sette, Sottano, Italy, 1937. 49

V. S. Shneyer, On the species-specificity of DNA: fifty years later, *Biochemistry (Mosc)*, 72(12):1377-84, 2007. DOI: 10.1134/S0006297907120127 50, 51, 52

J. De Ley, H. Cattoir, A. Reynaerts, The quantitative measurement of DNA hybridization from renaturation rates, *Biochem.*, 12, 133-142, 1970. DOI: 10.1111/j.1432-1033.1970.tb00830.x 50

L. Wayne, D. J. Brenner, R. R. Colwell, P. A. D. Grimont, O. Kandler, et al., Report of the Ad Hoc Committee on Reconciliation of Approaches to Bacterial Systematics, *Int J Syst Bacteriol*, 37:463-464, 1987. DOI: 10.1099/00207713-37-4-463 50

E. Stackebrandt and J. Ebers, Taxonomic parameters revisited: tarnished gold standards, *Microbiol. Today*, 33, 152-155, 2006. 50

M. Vulic, F. Dionisio, F. Taddei, M. Radman, "Molecular Keys to Speciation: DNA Polymorphism and the Control of Genetic Exchange in Enterobacteria", *Proc. Natl. Acad. Sci. USA*, 94, 9763-9767, 1997. DOI: 10.1073/pnas.94.18.9763 50

G. J. Olsen, C. R. Woese, R. Overbeek, The winds of (evolutionary) change: breathing new life into microbiology, *J. Bacteriol.*, 176:1-6, 1994. 50

E. Stackebrandt and B. M. Goebel, Taxonomic note: A place for DNA-DNA reassociation and 16S rRNA sequence analysis in the present species definition in bacteriology, *Int J Syst Bacteriol*, 44:846-849, 1994. DOI: 10.1099/00207713-44-4-846 50, 51

W. Ludwig, O. Strunk, S. Klugbauer, N. Klugbauer, M. Weizenegger, J. Neumaier, M. Bachleitner, K. H. Schleifer, Bacterial phylogeny based on comparative sequence analysis, *Electrophoresis*, 19:554-568, 1998. DOI: 10.1002/elps.1150190416 50

J. Goris, K. T. Konstantinidis, J. A. Klappenbach, T. Coenye, P. Vandamme, J. M. Tiedje, DNA-DNA hybridization values and their relationship to whole-genome sequence similarities, *Int. J. Syst. Evol. Microbiol.*, 57, 81-91, 2007. DOI: 10.1099/ijs.0.64483-0 51

D. Gevers, F. M. Cohan, J. G. Lawrence, B. G. Spratt, T. Coenye, E. J. Feil, E. Stackerbrandt, Y. van de Peer, P. Vandamme, F. L. Thompson, J. Swing, Opinion: Re-evaluating prokaryotic species, *Nat. Rev. Microbiol.*, 3, 733-739, 2005. DOI: 10.1038/nrmicro1236 49, 51, 73

T. P. Curtis and W. T. Sloan, Prokaryotic diversity and its limits: microbial community structure in nature and implications for microbial ecology, *Curr. Opin. Microbiol.*, 7, 221-226, 2004. DOI: 10.1016/j.mib.2004.04.010 51

S. J. Giovannoni and U. Stingl, Molecular diversity and ecology of microbial plankton, *Nature*, 437, 343-348, 2005. DOI: 10.1038/nature04158 51

W. P. Hanage, T. Kaijalainen, E. Herva, A. Saukkorupi, R. Syrjanen, B. G. Spratt, Using multilocus sequence data to define the pneumococcus, *J. Bacteriol.*, 187, 6223-6230, 2005. DOI: 10.1128/JB.187.17.6223-6230.2005 51

M. C. Maiden, J. A. Bygraves, E. Feil, G. Morelli, J. E. Russell, R. Urwin, Q. Zhang, J. Zhou, K. Zurth, D. A. Caugant, I. M. Feavers, M. Achtman, B. G. Spratt, Multilocus sequence typing: a portable approach to the identification of clones within populations of pathogenic microorganisms, *Proc. Natl. Acad. Sci. USA*, 95, 3140-3145, 1998. DOI: 10.1073/pnas.95.6.3140 51

Q. Wang, G. M. Garrity, J. M. Tiedje, J. R. Cole, Naive Bayesian classifier for rapid assignment of rRNA sequences into the new bacterial taxonomy, *Appl Environ Microbiol.*, 73(16):5261-7, 2007. DOI: 10.1128/AEM.00062-07 51, 59

J. R. Brown and W. F. Doolittle, Archaea and the prokaryote-to-eukaryote transition, *Microbiol Mol Biol Rev*, 61:456-502, 1997. 50

V. S. Shneer, *Bot. Zh.*, 76, 17-32, 1991. 51

A. A. Bannikova, Molecular markers and modern phylogenetics of mammals, *Zh. Obshch. Biol.*, 65, 278-305, 2004. 51

P. S. Soltis, D. E. Soltis, P. G. Wolf, D. L. Nickrent, S. M. Chaw, R. L. Chapman, The phylogeny of land plants inferred from 18S rDNA sequences: pushing the limits of rDNA signal?, *Mol Biol Evol*, 16:1774-1784, 1999. 51

D. L. Nickrent, M. A. Garcia, M. P. Martin, R. L. Mathiasen, A phylogeny of all species of Arceuthobium (Viscaceae) using nuclear and chloroplast DNA sequences, *Amer. J. Bot.*, 91, 125-138, 2004. DOI: 10.3732/ajb.91.1.125 51

C. R. Woese and G. E. Fox, Phylogenetic structure of the prokaryotic domain: The primary kingdoms, *Proc. Natl. Acad. Sci.*, 74(11):5088-90, 1977. DOI: 10.1073/pnas.74.11.5088 50

R. Floyd, A. Eyualem, A. Papert, M. Blaxter, Molecular barcodes for soil nematode identification, *Molec. Ecol.*, 11, 839-850, 2002. DOI: 10.1046/j.1365-294X.2002.01485.x 51

M. Markmann and D. Tautz, Reverse taxonomy: an approach towards determining the diversity of meiobenthic organisms based on ribosomal RNA signature sequences, *Philos. Trans. R. Soc. London B*, 360, 1917-1924, 2005. DOI: 10.1098/rstb.2005.1723 51

P. Greenhalgh and L. A. Steiner, Recombination activating gene 1 (Rag1) in zebrafish and shark, *Immunogenetics*, 41:54-55, 1995. DOI: 10.1007/BF00188438 51

P. D. N. Hebert, A. Cywinska, S. L. Ball, J. R. deWaard, Biological identifications through DNA barcodes, *Proc R Soc Lond B*, 270:313-321, 2003. DOI: 10.1098/rspb.2002.2218 52

M. Vences, M. Thomas, A. van der Meijden, Y. Chiari, D. R. Vieites, Comparative performance of the 16S rRNA gene in DNA barcoding of amphibians, *Front Zool*, 2:5-16, 2005b. DOI: 10.1186/1742-9994-2-5

K. F. Armstrong and S. L. Ball, DNA barcodes for biosecurity: invasive species identification, *philos. Trans. R. Soc. London B*, 360, 1813-1823, 2005. DOI: 10.1098/rstb.2005.2001 52

M. Vences, M. Thomas, R. M. Bonett, D. R. Vieites, Deciphering amphibian diversity through DNA barcoding: chances and challenges, *philos. Trans. R. Soc. London B*, 360, 1859-1868, 2005c. DOI: 10.1098/rstb.2005.1717 52

B. Gemeinholzer, C. Oberprieler, K. Bachmann, Screening the applicability of molecular markers for plant identification using the ITS 1 sequences of Asteraceae species belonging to the tribes Lactuceae and Anthemideae, *Taxon*, 55, 173-187, 2006. DOI: 10.2307/25065539 52

G. N. Feliner and J. A. Rossello, Better the devil you know? Guidelines for insightful utilization of nrDNA ITS in species-level evolutionary studies in plants, *Mol. Phylogenet. Evol.*, 44, 911-919, 2007. DOI: 10.1016/j.ympev.2007.01.013 52

M. W. Chase, N. Salamin, M. Wilkinson, J. M. Dunwell, R. P. Kesanakurthi, N. Haidar, V. Savolainen, Land plants and DNA barcodes: short-term and long-term goals, *Philos. Trans. R. Soc. London B*, 360, 1889-1895, 2005. DOI: 10.1098/rstb.2005.1720 52

J. A. Eisen and C. M. Fraser, Phylogenomics: intersection of evolution and genomics, *Science*, 300:1706-1707, 2003. DOI: 10.1126/science.1086292 52

K. Konstantinidis and J. M. Tiedje, Genomic insights that advance the species definition for prokaryotes, *Proc Natl Acad Sci USA*, 102:2567-2592, 2005. DOI: 10.1073/pnas.0409727102 52

M. Richter and R. Rossello-Mora, Shifting the genomic gold standard for the prokaryotic species definition, *Proc Natl Acad Sci USA*, 10;106(45):19126-31, 2009. DOI: 10.1073/pnas.0906412106 52, 54

M. Deloger, M. El Karoui, M-A. Petit, A genomic distance based on MUM indicates discontinuity between most bacterial species and genera, *J Bacteriol*, 191:91-99, 2009. DOI: 10.1128/JB.01202-08 52

E. Haywood-Farmer, S. P. Otto, The evolution of genomic base composition in bacteria, *Evolution*, 57(8):1783-92, 2003. DOI: 10.1554/01-535 53

G. M. Pupo, R. Lan, P. R. Reeves, Multiple independent origins of Shigella clones of Escherichia coli and convergent evolution of many of their characteristics, *Proc Natl Acad Sci USA*, 97(19):10567-10572, 2000. DOI: 10.1073/pnas.180094797 53

M. W. van Passel, E. E. Kuramae, A. C. Luyf, A. Bart, T. Boekhout, The reach of the genome signature in prokaryotes, *MC Evol Biol.*, 13;6:84, 2006. DOI: 10.1186/1471-2148-6-84 53

E. V. Koonin, L. Aravind, A. S. Kondrashov, The impact of comparative genomics on our understanding of evolution, *Cell*, 101: 573-576, 2000. DOI: 10.1016/S0092-8674(00)80867-3 52

D. T. Pride, R. J. Meinersmann, T. M. Wassenaar, M. J. Blaser, Evolutionary implications of microbial genome tetranucleotide frequency biases, *Genome Res.,* 13(2):145-58, 2003. DOI: 10.1101/gr.335003 54

J. Parkhill, B. W. Wren, K. Mungall, J. M. Ketley, C. Churcher, D. Basham, T. Chillingworth, R. M. Davies, T. Feltwell, S. Holroyd, et al., The genome sequence of the food-borne pathogen Campylobacter jejuni reveals hypervariable sequences, *Nature,* 403: 665-668, 2000. DOI: 10.1038/35001088 54

J.-F. Tomb, O. White, A. R. Kervalage, R. A. Clayton, G. G. Sutton, R. D. Fleischman, K. A. Ketchum, H. P. Klenk, S. Gill, B. A. Dougherty, et al., The complete genome sequence of the gastric pathogen Helicobacter pylori, *Nature,* 388: 539-547, 1997. DOI: 10.1038/41483 54

T. Coenye and P. Vandamme, Use of the genome signature in bacterial classification and identification, *Systematic and Applied Microbiology,* 27:175-185, 2004. DOI: 10.1078/072320204322881790 54

G. W. Stuart, K. Moffet, J. J. Leader, A comprehensive vertebrate phylogeny using vector representations of protein sequences from whole genomes, *Mol Biol Evol,* 19:554-562, 2002. 54

X. Xu, A. Janke, U. Arnason, The complete mitochondrial DNA sequence of the greater Indian rhinoceros, Rhinoceros unicornis, and the Phylogenetic relationship among Carnivora, Perissodactyla, and Artiodactyla (+ Cetacea), *Mol Biol Evol,* 13(9):1167-73, 1996. 54

A. Janke, X. Xu, U. Arnason, The complete mitochondrial genome of the wallaroo (Macropus robustus) and the phylogenetic relationship among Monotremata, Marsupialia, and Eutheria, *Proc Natl Acad Sci USA,* 94(4):1276-81, 1997. DOI: 10.1073/pnas.94.4.1276 54

U. Arnason, A. Gullberg, S. Gretarsdottir, B. Ursing, A. Janke, The mitochondrial genome of the sperm whale and a new molecular reference for estimating eutherian divergence dates, *J Mol Evol.,* 50(6):569-78, 2000. 54

M. Nikaido, M. Harada, Y. Cao, M. Hasegawa, N. Okada, Monophyletic origin of the order chiroptera and its phylogenetic position among mammalia, as inferred from the complete sequence of the mitochondrial DNA of a Japanese megabat, the Ryukyu flying fox (Pteropus dasymallus), *J Mol Evol.,* 51(4):318-28, 2000. DOI: 10.1007/s002390010094 55

A. Reyes, C. Gissi, G. Pesole, F. M. Catzeflis, C. Saccone, Where do rodents fit? Evidence from the complete mitochondrial genome of Sciurus vulgaris, *Mol Biol Evol.,* 17(6):979-83, 2000. 55

Z. G. Yu , X. W. Zhan , G. S. Han, R. W. Wang, V. Anh, K. H. Chu, Proper distance metrics for phylogenetic analysis using complete genomes without sequence alignment, *Int J Mol Sci.,* 11(3):1141-54, 2010. DOI: 10.3390/ijms11031141 55

Z.-G. Yu, L.-Q. Zhou, V. Anh, K. H. Chu, S.-C. Long, J.-Q. Deng, Phylogeny of prokaryotes and chloroplasts revealed by a simple composition approach on all protein sequences from whole genome without sequence alignment, *J. Mol. Evol.*, 60, 538-545, 2005. DOI: 10.1007/s00239-004-0255-9 55

R. L. Charlebois, R. G. Beiko, M. A. Ragan, Branching out, *Nature*, 421:217-217, 2003. DOI: 10.1038/421217a 55

M. Li, J. H. Badger, X. Chen, S. Kwong, P. Kearney, H. Zhang, An information-based sequence distance and its application to whole mitochondrial genome phylogeny, *Bioinformatics*, 17, 149-1547, 2001. DOI: 10.1093/bioinformatics/17.2.149 55

Z. Dawy, J. Hagenauer, P. Hanus and J. C. Mueller, Mutual information based distance measures for classification and content recognition with applications to genetics, *Communications, 2005, ICC 2005. 2005 IEEE international conference on*, 2:820-824, 2005. 55

A. Apostolico, M. Comin, L. Parida, Mining, compressing and classifying with extensible motifs, *Algorithms for Molecular Biology*, 1, 4, 2006. DOI: 10.1186/1748-7188-1-4 55

Z. G. Yu, Z. Mao, L. Q. Zhou, V. V. Anh, A mutual information based sequence distance for vertebrate phylogeny using complete mitochondrial genomes, *In Proceeding of the 3nd International Conference on Natural Computation (ICNC2007)*, Haikou, China, pp. 253-257, August 2007. DOI: 10.1109/ICNC.2007.78 56

A. Lempel and J. Ziv, On the complexity of finite sequences, *IEEE Trans. Inf. Theory*, 22, 75-88, 1976. DOI: 10.1109/TIT.1976.1055501 56

H. H. Otu and K. Sayood, A new sequence distance measure for phylogenetic tree construction, *Bioinformatics*, 19(16), 2003. DOI: 10.1093/bioinformatics/btg295 57

M. Takahashi, K. Kryukov, N. Saitou, Estimation of bacterial species phylogeny through oligonucleotide frequency distances, *Genomics*, 93(6):525-33, 2009. DOI: 10.1016/j.ygeno.2009.01.009 59

C. F. Davenport and B. Tümmler, Abundant oligonucleotides common to most bacteria, *PLoS One*, 23;5(3):e9841, 2010. DOI: 10.1371/journal.pone.0009841 60

N. R. Pace, A molecular view of microbial diversity and the biosphere, *Science*, 276: 734740, 1997. DOI: 10.1126/science.276.5313.734 61

M. S. Rappe and S. J. Giovannoni, The uncultured microbial majority, *Annu Rev Microbiol*, 57: 369394, 2003. DOI: 10.1146/annurev.micro.57.030502.090759 61

R. D. Fleischmann, M. D. Adams, O. White, R. A. Clayton, E. F. Kirkness, et al., Whole-genome random sequencing and assembly of haemophilus influenzae rd., *Science*, 269: 496512, 1995. DOI: 10.1126/science.7542800 61

100   BIBLIOGRAPHY

The dawning of a new microbial age, *in The New Science of Metagenomics: Revealing the Secrets of Our Microbial Planet*, p. 2, The National Academies Press, Washington, DC, 2007. 61

M. Achtman, and M. Wagner, Microbial diversity and the genetic nature of microbial species, *Nat. Rev. Microbiol.*, 6:431-440, 2008. 61

P. Hugenholtz, Exploring prokaryotic diversity in the genomic era, *Genome Biol.*, 3:REVIEWS0003, 2002. DOI: 10.1186/gb-2002-3-2-reviews0003 61

E. DeLong, Microbial Community Genomics in the Ocean, *Nature Reviews*, 3:459469, 2005. DOI: 10.1038/nrmicro1158 61

E. DeLong, C. Preston, T. Mincer, V. Rich, S. Hallam, N. U. Frigaard, A. Martinez, M. Sullivan, R. Edwards, B. Brito, S. Chisholm, D. Karl, Community Genomics Among Stratified Microbial Assemblages in the Oceans Interior, *Science*, 311:496503, 2006. DOI: 10.1126/science.1120250 61

R. Daniel, The Metagenomics of Soil, *Nature rev*, 3:470-478, 2005 DOI: 10.1038/nrmicro1160 61

S. M. Barns, R. E. Fundyga, M. W. Jeffries, N. R. Pace, Remarkable archeal diversity detected in a Yellowstone National Park hot spring environment, *Proc Natl Acad Sci USA*, 91:16091613, 1994. DOI: 10.1073/pnas.91.5.1609 61

R. Huber, H. Huber, K. O. Stetter, Towards ecology of hyperthermophiles: biotypes, new isolation strategies and novel metabolic properties, *FEMS Microbiol Rev*, 24:615623, 2002. 61

B. C. Christner, B. H. Kvitko, J. N. Reeve, Molecular identification of Bacteria and Eukraya inhabiting an Antarctic cryoconite hole, *Extremophiles*, 7:177183, 2003. DOI: 10.1007/s00792-002-0309-0 61

S. Bellnoch, Prokaryotic genetic diversity throughout the salinity gradient of a coastal solar saltern, *Environmental Microbiology*, 4:349360, 2002. DOI: 10.1046/j.1462-2920.2002.00306.x 61

P. Lorenz and J. Eck, Metagenomics and industrial applications, *Nature Reviews*, 3:510516, 2005. DOI: 10.1038/nrmicro1161 62

C. Schmeisser, H. Steele, W. Streit, Metagenomics, biotechnology with nonculturable microbes, *Appl Microbiol Biotechnol*, 75:955962, 2007. DOI: 10.1007/s00253-007-0945-5 62

P. J. Turnbaugh, R. E. Ley, M. Hamady, C. M. Fraser-Liggett, R. Knight, J. I. Gordon, The Human Microbiome Project, *Nature*, 449:804810, 2007. DOI: 10.1038/nature06244 62

J. P. Noonan, G. Coop, S. Kudaravalli, D. Smith, J. Krause, J. Alessi, F. Chen, D. Platt, S. Paabo, J. K. Pritchard, E. M. Rubin, Sequencing and Analysis of Neanderthal Genomic DNA, *Science*, 314 (5802):11131118, 2006. DOI: 10.1126/science.1131412 62

R. E. Green, J. Krause, S. E. Ptak, A. W. Briggs, M. T. Ronan, et al., Analysis of one million base pairs of Neanderthal DNA, *Nature*, 444:330336, 2006. DOI: 10.1038/nature05336 62

A. McHardy and I. Rigoutsos, Whats in the mix: phylogenetic classifcation of metagenome sequence samples, *Current Opinion in Microbiology*, 10:499503, 2007. DOI: 10.1016/j.mib.2007.08.004 62

A. Andersson, M. Lindberg, H. Jakobsson, F. Backhed, P. Nyren, L. Engstrand, Comparative Analysis of Human Gut Microbiota by Barcoded Pyrosequencing, *PLoS One*, 3(7):e2836, 2008. DOI: 10.1371/journal.pone.0002836 62

S. Tringe, C. von Mering, A. Kobayashi, A. Salamov, K. Chen, H. Chang, M. Podar, J. Short, E. Mathur, J. Detter, P. Bork, P. Hugenholtz, E. Rubin, Comparative Metagenomics of Microbial Communities, *Science*, 308:5547, 2005. DOI: 10.1126/science.1107851 62, 63

W. F. Doolittle, Phylogenetic classification and the universal tree, *Science*, 284:2124-2129, 1999. DOI: 10.1126/science.284.5423.2124 62, 73

J. C. Wooley, A. Godzik, I. Friedberg, A primer on metagenomics, *PLoS Comput Biol*, 26;6(2):e1000667, 2010. DOI: 10.1371/journal.pcbi.1000667 62, 63

D. Willner, M. Furlan, M. Haynes, R. Schmieder, F. E. Angly, et al., Metagenomic analysis of respiratory tract DNA viral communities in cystic fibrosis and non-cystic fibrosis individuals, *PLoS ONE*, 4: e7370, 2009. DOI: 10.1371/journal.pone.0007370 62

G.W. Tyson, J. Chapman, P. Hugenholtz, E.E. Allen, R.J. Ram, P.M. Richardson, V.V. Solovyev, E.M. Rubin, D.S. Rokhsar, and J.F. Banfield, Community structure and metabolism through reconstruction of microbial genomes from the environment, *Nature*, 428:37-43, 2004. DOI: 10.1038/nature02340 62

H. Garcia Martin, N. Ivanova, V. Kunin, F.Warnecke, K.W. Barry, A.C. McHardy, C. Yeates, S. He, A.A. Salamov, E. Szeto, E. Dalin, N.H. Putnam, H.J. Shapiro, J.L. Pangilinan, I. Rigoutsos, N.C. Kyrpides, L.L. Blackall, K.D. McMahon, and P. Hugenholtz, Metagenomic analysis of two enhanced biological phosphorus removal (EBPR) sludge communities, *Nat. Biotechnol*, 24:1263-1269, 2006. DOI: 10.1038/nbt1247 62

M. Strous, E. Pelletier, S. Mangenot, T. Rattei, A. Lehner, M.W. Taylor, M. Horn, H. Daims, D. Bartol-Mavel, P. Wincker, V. Barbe, N. Fonknechten, D. Vallenet, B. Segurens, C. Schenowitz-Truong, C. Medigue, A. Collingro, B. Snel, B.E. Dutilh, H.J. Op den Camp, C. van der Drift, I. Cirpus, K.T. van de Pas-Schoonen, H.R. Harhangi, L. van Niftrik, M. Schmid, J. Keltjens, J. van de Vossenberg, B. Kartal, H. Meier, D. Frishman, M.A. Huynen, H.W. Mewes, J. Weissenbach, M.S. Jetten, M. Wagner, and D. Le Paslier, Deciphering the evolution and metabolism of an anammox bacterium from a community genome, *Nature*, 440:790-794, 2006. DOI: 10.1038/nature04647 62

T. Woyke, H. Teeling, N. N. Ivanova, M. Huntemann, M. Richter, F. O. Gloeckner, D. Boffelli, I. J. Anderson, K. W. Barry, H. J. Shapiro, E. Szeto, N. C. Kyrpides, M. Mussmann, R. Amann, C. Bergin, C. Ruehland, E. M. Rubin, and N. Dubilier, Symbiosis insights through metagenomic analysis of a microbial consortium, *Nature*, 443:950-955, 2006. DOI: 10.1038/nature05192 62

F. Warnecke, P. Luginbuhl, N. Ivanova, M. Ghassemian, T. H. Richardson, J. T. Stege, M. Cayouette, A. C. McHardy, G. Djordjevic, N. Aboushadi, R. Sorek, S. G. Tringe, M. Podar, H. G. Martin, V. Kunin, D. Dalevi, J. Madejska, E. Kirton, D. Platt, E. Szeto, A. Salamov, K. Barry, N. Mikhailova, N. C. Kyrpides, E. G. Matson, E. A. Ottesen, X. Zhang, M. Hernandez, C. Murillo, L. G. Acosta, I. Rigoutsos, G. Tamayo, B. D. Green, C. Chang, E. M. Rubin, E. J. Mathur, D. E. Robertson, P. Hugenholtz, J. R. Leadbetter, Metagenomic and functional analysis of hindgut microbiota of a wood-feeding higher termite, *Nature*, 50:560-565, 2007. DOI: 10.1038/nature06269 63

F. Sanger, and A. R. Coulson, A rapid method for determining sequences in DNA by primed synthesis with DNA polymerase, *J. Mol. Biol.*, 94:441-448, 1975. DOI: 10.1016/0022-2836(75)90213-2 63

F. Sanger, S. Nicklen, A. R. Coulson, DNA sequencing with chain-terminating inhibitors, *Proc. Natl. Acad. Sci. USA*, 74:5463-5467, 1977. DOI: 10.1073/pnas.74.12.5463 63

Joint Genome Institute, *http://www.jgi.doe.gov/CSP/index.html.* 63

E. K. Wommack, J. Bhavsar, J. Ravel, Metagenomics: Read length matters, *Appl Environ Microbiol.*, 74(5):1453-1463, 2008. DOI: 10.1128/AEM.02181-07 63

E. S. Lander and M. S. Waterman, Genomic mapping by fingerprinting random clones: a mathematical analysis, *Genomics*, 2: 231-239, 1988. DOI: 10.1016/0888-7543(88)90007-9 63

E. R. Mardis, Next-generation DNA sequencing methods, *Annu Rev Genomics Hum Genet.*, 9:387-402, 2008. DOI: 10.1146/annurev.genom.9.081307.164359 63

M. L. Metzker, Sequencing technologies - the next generation, *Nat Rev Genet.*, 11(1):31-46, 2010. DOI: 10.1038/nrg2626 63

S. Fox, S. Filichkin, T. C. Mockler, Applications of ultra-high-throughput sequencing, *Methods Mol Biol.*, 553:79-108, 2009. DOI: 10.1007/978-1-60327-563-7_5 63

http://www.454.com/products-solutions/system-features.asp 63

http://www.illumina.com/downloads/SQGAIIxspecsheet20409LR.pdf 63

http://www.appliedbiosystems.com/ABHome/applicationstechnologies/SOLiDSystemSequencing/ 63

http://www.helicosbio.com/Technology/TrueSingleMoleculeSequencing/tSMStradePerformance/ 63

T. Z. DeSantis, P. Hugenholtz, N. Larsen, M. Rojas, E. L. Brodie, K. Keller, T. Huber, D. Dalevi, P. Hu, and G. L. Andersen. Greengenes, a chimera-checked 16S rRNA gene database and workbench compatible with ARB, *Appl. Environ. Microbiol.*, 72:5069-5072, 2006. DOI: 10.1128/AEM.03006-05 63

R. J. Case , Y. Boucher, I. Dahllof, C. Holmstrom, F. W. Doolittle, et al., Use of 16srRNA and rpob genes as molecular markers for microbial ecology studies, *Appl Environ Microbiol,* 73: 278-288, 2007. DOI: 10.1128/AEM.01177-06 63

C. von Mering, P. Hugenholtz, J. Raes, S. G. Tringe, T. Doerks, L. J. Jensen, N. Ward, P. Bork, Quantitative phylogenetic assessment of microbial communities in diverse environments, *Science,* 315:1126-1130, 2007. DOI: 10.1126/science.1133420 63

E. Mahenthiralingam, A. Baldwin, P. Drevinek, E. Vanlaere, P. Vandamme, et al., Multilocus sequence typing breathes life into a microbial metagenome, *PLoS ONE,* 1: e17. doi:10.1371, 2006. DOI: 10.1371/journal.pone.0000017 63

M. T. Suzuki and S. J. Giovannoni, Bias caused by template annealing in the amplification of mixtures of 16S rRNA genes by PCR, *Appl. Environ. Microbiol.*, 62:625-630, 1996. 64

F. von Wintzingerode, U. B. Gobel, E. Stackebrandt, Determination of microbial diversity in environmental samples: pitfalls of PCR based rRNA analysis, *FEMS Microbiol. Rev.,* 21:213-229, 1997. DOI: 10.1111/j.1574-6976.1997.tb00351.x 64

M. Wu and J. A. Eisen, A simple, fast, and accurate method of phylogenomic inference, *Genome Biol,* 9(10):R151, 2008. DOI: 10.1186/gb-2008-9-10-r151 64

M. Stark, S. A. Berger, A. Stamatakis, C. von Mering, MLTreeMap - accurate Maximum Likelihood placement of environmental DNA sequences into taxonomic and functional reference phylogenies, *BMC Genomics,* 11:461, 2010. DOI: 10.1186/1471-2164-11-461 64

S. L. Salzberg and J. A. Yorke, Beware of mis-assembled genomes, *Bioinformatics,* 21:4320-4321, 2005. DOI: 10.1093/bioinformatics/bti769 64

P. L. Johnson and M. Slatkin, Inference of population genetic parameters in metagenomics: a clean look at messy data, *Genome Res,* 16:1320-1327, 2006. DOI: 10.1101/gr.5431206 64

P. Green, *www.phrap.org.* DOI: 10.1088/1126-6708/2007/04/025 64

X. Huang and A. Madan, CAP3: A DNA sequence assembly program, *Genome Res,* 9:868-77, 1999. DOI: 10.1101/gr.9.9.868 64

S. Batzoglou, D. B. Jaffe, K. Stanley, J. Butler, S. Gnerre et al., ARACHNE: a whole-genome shotgun assembler, *Genome Res,* 12: 177-189, 2002. DOI: 10.1101/gr.208902 64

D. B. Jaffe, J. Butler, S. Gnerre, E. Mauceli, K. Lindblad-Toh, J. P. Mesirov, M. C. Zody, E. S. Lander, Whole-genome sequence assembly for mammalian genomes: Arachne 2, *Genome Res*, 13: 91-96, 2003. DOI: 10.1101/gr.828403 64

S. Aparicio, J. Chapman, E. Stupka, N. Putnam, J. ming Chia, et al., Wholegenome shotgun assembly and analysis of the genome of fugu rubripes, *Science*, 297: 1301-1310, 2002. DOI: 10.1126/science.1072104 65

E. W. Myers, G. G. Sutton, A. L. Delcher, I. M. Dew, D. P. Fasulo, et al., A wholegenome assembly of drosophila, *Science*, 287: 2196-2204, 2000. DOI: 10.1126/science.287.5461.2196 65

P. A. Pevzner, H. Tang, M. S. Waterman, An eulerian path approach to DNA fragment assembly, *Proc Natl Acad Sci USA*, 98: 9748-9753, 2001. DOI: 10.1073/pnas.171285098 65

M. J. Chaisson, and P. A. Pevzner, Short read fragment assembly of bacterial genomes, *Genome Res.*, 18: 324-330, 2008. DOI: 10.1101/gr.7088808 65

M. Pop, Genome assembly reborn: recent computational challenges, *Bioinformatics*, 4: 354- 366, 2009. DOI: 10.1093/bib/bbp026 65

R. L. Warren, G. G. Sutton, S. J. Jones, R. A. Holt, Assembling millions of short DNA sequences using SSAKE, *Bioinformatics*, 23: 500-501, 2007. DOI: 10.1093/bioinformatics/btl629 65

W. R. Jeck, J. A. Reinhardt, D. A. Baltrus, M. T. Hickenbotham, V. Magrini, E. R. Mardis, J. L. Dangl, and C. D. Jones, Extending assembly of short DNA sequences to handle error, *Bioinformatics*, 23: 2942-2944, 2007. DOI: 10.1093/bioinformatics/btm451 65

J. C. Dohm, C. Lottaz, T. Borodina, H. Himmelbauer, SHARCGS, a fast and highly accurate short-read assembly algorithm for de novo genomic sequencing, *Genome Res.*, 17: 1697-1706, 2007. DOI: 10.1101/gr.6435207 65

D. R. Zerbino and E. Birney, Velvet: Algorithms for de novo short read assembly using de Bruijn graphs, *Genome Res.*, 18: 821-829, 2008. DOI: 10.1101/gr.074492.107 65

L. Krause et al., Phylogenetic classification of short environmental DNA fragments, *Nucleic Acids Res*, 36, 2230-2239, 2008. DOI: 10.1093/nar/gkn038 65, 66

D. H. Huson, A. F. Auch, J. Qi, S. C. Schuster, MEGAN analysis of metagenomic data, *Genome Res.*, 17, 377-386, 2007. DOI: 10.1101/gr.5969107 66

D. H. Huson, D. C. Richter, S. Mitra, A. F. Auch, S. C. Schuster, Methods for comparative metagenomics *BMC Bioinformatics.*, 10(Suppl 1):S12, 2009. DOI: 10.1186/1471-2105-10-S1-S12 66

S. F. Altschul, Gapped BLAST and PSI-BLAST: a new generation of protein database search program, *Nucleic Acids Res.*, 125:3389-3402, 1997. DOI: 10.1093/nar/25.17.3389 66

E.A. Dinsdale, Microbial ecology of four coral atolls in the Northern Line Islands, *PLoS One*, 3:e1584, 2008. DOI: 10.1371/journal.pone.0001584 66

H. M. Monzoorul, S. Tarini, K. Dinakar, S. M. Sharmila, SOrt-ITEMS: sequence orthology based approach for improved taxonomic estimation of metagenomic sequences, *Bioinformatics*, 25:1722-1730, 2009. DOI: 10.1093/bioinformatics/btp317 66

A. Brady and S. L. Salzberg, Phymm and PhymmBL: metagenomic phylogenetic classification with interpolated markov models, *Nat Methods*, 6: 673-676, 2009. DOI: 10.1038/nmeth.1358 66, 67

A. C. McHardy, H. G. Martin, A. Tsirigos, P. Hugenholtz, I. Rigoutsos, Accurate phylogenetic classification of variable-length DNA fragments, *Nat Methods*, 4:63-72, 2007. DOI: 10.1038/nmeth976 67

N. N. Diaz, L. Krause, A. Goesmann, K. Niehaus, T. W. Nattkemper, TACOA: taxonomic classification of environmental genomic fragments using a kernelized nearest neighbor approach, *BMC Bioinformatics*, 10:56, 2009. DOI: 10.1186/1471-2105-10-56 67

T. Kohonen, Self-organized formation of topologically correct feature maps, *Biol. Cybern.*, 43: 5969, 1982. 68

T. Kohonen, E. Oja, O. Simula, A. Visa, J. Kangas, Engineering applications of the self-organizing map, *Proc. IEEE*, 84: 13581384, 1996. DOI: 10.1109/5.537105 68

G. J. Dick, et al., Community-wide analysis of microbial genome sequence signatures, *Genome Biol*, 10:R85, 2009. DOI: 10.1186/gb-2009-10-8-r85 68

K. U. Foerstner, C. von Mering, S. D. Hooper, P. Bork, Environments shape the nucleotide composition of genomes, *EMBO Rep*, 6:1208-1213, 2005. DOI: 10.1038/sj.embor.7400538 68, 79

H. Willenbrock, C. Friis, A. S. Juncker, D. W. Ussery, An environmental signature for 323 microbial genomes based on codon adaptation indices, *Genome Biol*, 7:R114, 2006. DOI: 10.1186/gb-2006-7-12-r114 68

J. Raes, K. U. Foerstner, P. Bork, Get the most out of your metagenome: computational analysis of environmental sequence data, *Curr Opin Microbiol*, 10:490-498, 2007. DOI: 10.1016/j.mib.2007.09.001 68

S. Paul, S. K. Bag, S. Das, E. T. Harvill, C. Dutta, Molecular signature of hypersaline adaptation: insights from genome and proteome composition of halophilic prokaryotes, *Genome Biol*, 9:R70, 2008. DOI: 10.1186/gb-2008-9-4-r70 68

C. Martin, N. N. Diaz, J. Ontrup, T. W. Nattkemper, Hyperbolic SOMbased clustering of DNA fragment features for taxonomic visualization and classification, *Bioinformatics*, 24:1568-1574, 2008. DOI: 10.1093/bioinformatics/btn257 68

C. Chan, A. Hsu, S. Halgamuge, S. Tang, Binning sequences using very sparse labels within a metagenome, *BMC Bioinformatics*, 9:215, 2008b. DOI: 10.1186/1471-2105-9-215 68

S. Chatterji, I. Yamazaki, Z. Bai, J. Eisen, CompostBin: A DNA composition-based algorithm for binning environmental shotgun reads, *In Research in Computational Molecular Biology, 12th Annual International Conference, RECOMB 2008, Singapore, March 30 - April 2, 2008. Proceedings, Lecture Notes in Computer Science*, Volume 4955, Springer, 2008. 68

W. J. Kent, BLAT-the BLAST-like alignment tool, *Genome Res*, 12(4), 656-664 , 2002. DOI: 10.1101/gr.229202 68

A. Kislyuk, S. Bhatnagar, J. Dushoff, J. Weitz, Unsupervised statistical clustering of environmental shotgun sequences, *BMC Bioinformatics*, 10:316 , 2009. DOI: 10.1186/1471-2105-10-316 69

Y. W. Wu and Y. Ye, A Novel Abundance-Based Algorithm for Binning Metagenomic Sequences Using l-Tuples, *In Research in Computational Molecular Biology, of Lecture Notes in Computer Science.*, Volume 6044. Edited by Berger B. Springer DOI: 10.1007/978-3-642-12683-3_35 69

I. Sharon, A. Pati, V. M. Markowitz, et al. A statistical framework for the functional analysis of metagenomes, *In RECOMB 2009*, Springer Berlin / Heidelberg, Tucson, AZ, 496-511, 2009. DOI: 10.1007/978-3-642-02008-7_35

B. Yang, Y. Peng, H. C. Leung, S. M. Yiu, J. C. Chen, F. Y. Chin. Unsupervised binning of environmental genomic fragments based on an error robust selection of l-mers, *BMC Bioinformatics*, 16;11, 2010. DOI: 10.1186/1471-2105-11-S2-S5 69

D. R. Kelley and S. L. Salzberg, Clustering metagenomic sequences with interpolated Markov models, *BMC Bioinformatics*, 2;11:544, 2010. DOI: 10.1186/1471-2105-11-544 69

O. T. Avery, C. M. MacLeod, M. McCarty, Studies on the chemical nature of the substance inducing transformation of pneumococcal types. Inductions of transformation by a desoxyribonucleic acid fraction isolated from pneumococcus type III, *J. Exp. Med.*, 79, 137-158, 1944. DOI: 10.1084/jem.79.2.137 71

J. Hacker, L. Bender, M. Ott, J. Wingender, B. Lund, R. Marre, W. Goebel, Deletions of chromosomal regions coding for fimbriae and hemolysins occur in vitro and in vivo in various extraintestinal Escherichia coli isolates, *Microb Pathog*, 8:213-225, 1990. DOI: 10.1016/0882-4010(90)90048-U 71

U. Dobrindt, B. Hochhut, U. Hentschel, J. Hacker, Genomic islands in pathogenic and environmental microorganisms, *Nat Rev Microbiol*, 2(5):414-424, 2004. DOI: 10.1038/nrmicro884

S. Garcia-Vallve, A. Romeu, J. Palau, Horizontal gene transfer in bacterial and archaeal complete genomes, *Genome Res.*, 10:1719-2, 2000. DOI: 10.1101/gr.130000 72, 80

J. Hacker and J. B. Kaper, Pathogenicity islands and the evolution of microbes, *Annu Rev Microbiol,* 54:641-79, 2000. DOI: 10.1146/annurev.micro.54.1.641 72

U. Hentschel and J. Hacker, Pathogenicity islands: the tip of the iceberg, *Microbes Infect,* 3(7):545-8, 2001. DOI: 10.1016/S1286-4579(01)01410-1 72

E. Grohmann, G. Muth, M. Espinosa, Conjugative plasmid transfer in gram-positive bacteria, *Microbiol Mol Biol Rev,* 67, 277-301, 2003. DOI: 10.1128/MMBR.67.2.277-301.2003 72

N. D. Zinder and J. Lederberg, Genetic exchange in Salmonella, *J Bacteriol,* 64, 679-699, 1952. 72

H. Brussow, C. Canchaya, W. D. Hardt, Phages and the evolution of bacterial pathogens: from genomic rearrangements to lysogenic conversion, *Microbiol Mol Biol Rev,* 68: 560-602, 2004. DOI: 10.1128/MMBR.68.3.560-602.2004 72

D. Dubnau, DNA uptake in bacteria, *Annu Rev Microbiol,* 53, 217-244, 1999. DOI: 10.1146/annurev.micro.53.1.217 72

J. R. Zaneveld, D. R. Nemergut, R. Knight, Are all horizontal gene transfers created equal? Prospects for mechanism-based studies of HGT patterns, *Microbiology,* 154, 1-15, 2008. DOI: 10.1099/mic.0.2007/011833-0 72

E.V. Koonin, K. S. Makarova, L. Aravind, Horizontal gene transfer in prokaryotes: quantification and classification, *Annu Rev Microbiol,* 55:709-742, 2001. DOI: 10.1146/annurev.micro.55.1.709 72, 75

J. Maynard-Smith and N. H. Smith, Detecting recombination from gene trees, *Mol. Biol. Evol.,* 15, 590-599, 1998. 72

G. Lecointre, L. Rachdi, P. Darlu, E. Denamur, Escherichia coli molecular phylogeny using the incongruence length difference test, *Mol. Biol. Evol.,* 15, 1685-1695, 1998. 72

Y. Nakamura, T. Itoh, H. Matsuda, T. Gojobori, Biased biological functions of horizontally trans-ferred genes in prokaryotic genomes, *Nat Genet,* 36(7):760-766, 2004. DOI: 10.1038/ng1381 72, 73, 79

J. G. Lawrence and H. Ochman, Amelioration of bacterial genomes: rates of change and exchange, *J Mol Evol,* 44(4):383-397, 1997. DOI: 10.1007/PL00006158 72, 74, 76, 79, 82

A. B. Simonson, J. A. Servin, R. G. Skophammer, C. W. Herbold, M. C. Rivera, J. A. Lake, Decoding the genomic tree of life, *Proc Natl Acd Sci USA,* 102: 6608-6613, 2005. DOI: 10.1073/pnas.0501996102 72, 73, 74

J. G. Lawrencee, Gene transfer, speciation, and the evolution of bacterial geneomes, *Curr. Opin. Microbiol.,* 2: 519-523, 1999. DOI: 10.1016/S1369-5274(99)00010-7 73

B. Snel, P. Bork, M. A. Huynen, Genomes in flux: the evolution of archaeal and proteobacterial gene content, *Genome Res.*, 12(1):17-25, 2002. DOI: 10.1101/gr.176501 73

E. Bapteste, E. Susko, J. Leigh, D. MacLeod, R. L. Charlebois, W. F. Doolittle, Do orthologous gene phylogenies really support treethinking?, *BMC Evol Biol*, 5:33, 2005. DOI: 10.1186/1471-2148-5-33 73

T. M. Embley and W. Martin, Eukaryotic evolution, changes and challenges, *Nature*, 440:623-630, 2006. DOI: 10.1038/nature04546 73

H. Ochman, E. Lerat, V. Daubin, Examining bacterial species under the specter of gene transfer and exchange, *Proc Natl Acad Sci USA*, 102(Suppl 1):6595-6599, 2005. DOI: 10.1073/pnas.0502035102 73

F. Ge, L. S. Wang, J. Kim, The cobweb of life revealed by genomescale estimates of horizontal gene transfer, *PLoS Biol*, 3:e316, 2005. DOI: 10.1371/journal.pbio.0030316 73

F. Darwin, *The Life and Letters of Charles Darwin*, John Murray, London, 1887. 73

F. de la Cruz and J. Davies, Horizontal gene transfer and the origin of species: lessons from bacteria, *Trends Microbiol.*, 8, 128-132, 2000. DOI: 10.1016/S0966-842X(00)01703-0 73

L. Aravind, R. L. Tatusov, Y. I. Wolf, D. R. Walker, E. V. Koonin, Evidence for massive gene exchange between archaeal and bacterial hyperthermophiles, *Trends Genet.*, 14, 442-444, 1998. DOI: 10.1016/S0168-9525(98)01553-4 74

K. E. Nelson, R. A. Clayton, S. R. Gill, M. L. Gwinn, R. J. Dodson, et al., Evidence for lateral gene transfer between Archaea and bacteria from genome sequence of Thermotoga maritima, *Nature*, 399, 323-329, 1999. DOI: 10.1038/20601 74

L. M. Van Blerkom, Role of viruses in human evolution, *Am J Phys Anthropol*, Suppl 37: 14-46, 2003. DOI: 10.1002/ajpa.10384 74

D. J. Hedges and M. A. Batzer, From the margins of the genome: mobile elements shape primate evolution, *Bioessays*, 27(8):785-94, 2005. DOI: 10.1002/bies.20268 74

J. C. Dunning Hotopp et al., Widespread Lateral Gene Transfer from Intracellular Bacteria to Multicellular Eukaryotes, *Science*, 317:1753-6, 2007. DOI: 10.1126/science.1142490 74

W. F. Doolittle, Y. Boucher, C. L. Nesbo, C. J. Douady, J. O. Andersson, A. J. Roger, How big is the iceberg of which organellar genes in nuclear genomes are but the tip?, *Philos Trans R Soc Lond B Biol Sci*, 358(1429):39-57, discussion 57-8, 2003. DOI: 10.1098/rstb.2002.1185 74

R. Bock, J. N. Timmis, Reconstructing evolution: gene transfer from plastids to the nucleus, *Bioessays*, 30(6):556-66, 2008. DOI: 10.1002/bies.20761 74

J. M. Burke and M. L. Arnold, Genetics and the fitness of hybrids, *Annu Rev Genet*, 35:31-52, 2001. DOI: 10.1146/annurev.genet.35.102401.085719 74

J. C. Silva and M. G. Kidwell, Horizontal transfer and selection in the evolution of P elements, *Mol Biol Evol*, 17:1542-57, 2000. 74

H. Won, S. Renner, Horizontal gene transfer from flowering plant to Gnetum, *Proc Natl Acad Sci USA*, 100:10824-29, 2003. DOI: 10.1073/pnas.1833775100 74

P. Veronico, J. Jones, M. Di Vito, C. De Giorgi, Horizontal transfer of a bacterial gene involved in polyglutamate biosynthesis to the plant parasitic nemathod Meloidogyne artiellia, *FEBS Letts*, 508:470-4, 2001. DOI: 10.1016/S0014-5793(01)03132-5 74

M. C. Intieri and M. Buatti, The horizontal transfer of Agrobacterium rhizogenes genes and the evolution of the genus Nicotiana, *Mol Phyl Evol*, 20:100-10, 2001. DOI: 10.1006/mpev.2001.0927 74

L. Mallet, J. Becq, P. Deschavanne, Whole genome evaluation of horizontal transfers in the pathogenic fungus Aspergillus fumigatus, *BMC Genomics*, 11:171, 2010. DOI: 10.1186/1471-2164-11-171 74, 77

R. Jain, M. C. Rivera, J. A. Lake, Horizontal gene transfer among genomes: The complexity hypothesis, *Proc Natl Acad Sci USA*, 96:3801-3806, 1999. DOI: 10.1073/pnas.96.7.3801 73

S. H. Yoon, C. G. Hur, H. Y. Kang, Y. H. Kim, T. K. Oh, J. F. Kim, A computational approach for identifying pathogenicity islands in prokaryotic genomes, *BMC Bioinformatics*, 6: 184, 2005. DOI: 10.1186/1471-2105-6-184 75

K. V. Srividhya, G. V. Rao, L. Raghavenderan, P. Mehta, J. Prilusky, S. Manicka, J. L. Sussman, S. Krishnaswamy, Database and Comparative Identification of prophages, *LNCIS*, 344: 863-868, 2006. DOI: 10.1007/978-3-540-37256-1_110 75

S. R. Santos and H. Ochman, Identification and phylogenetic sorting of bacterial lineages with universally conserved genes and proteins, *Environ Microbiol*, 6:754-759, 2004. DOI: 10.1111/j.1462-2920.2004.00617.x 75

H. Y. Ou, X. He, E. M. Harrison, B. R. Kulasekara, A. B. Thani, A. Kadioglu, S. Lory, J. C. Hinton, M. R. Barer, Z. Deng, K. Rajakumar, Mobilome-FINDER: web-based tools for in silico and experimental discovery of bacterial genomic islands, *Nucleic Acids Res*, 35:W97-W104, 2007. DOI: 10.1093/nar/gkm380 75

H. Chiapello, I. Bourgait, F. Sourivong, G. Heuclin, A. Gendrault-Jacquemard, M. A. Petit, M. El Karoui, Systematic determination of the mosaic structure of bacterial genomes: species backbone versus strain-specific loops, *BMC Bioinformatics*, 6:171, 2005. DOI: 10.1186/1471-2105-6-171 75

M. G. Langille, W. W. Hsiao, F. S. Brinkman, Evaluation of genomic island predictors using a comparative genomics approach, *BMC Bioinformatics*, 9:329, 2008. DOI: 10.1186/1471-2105-9-329 75, 79

S. Podell and T. Gaasterland, DarkHorse: a method for genome-wide prediction of horizontal gene transfer, *Genome Biol*, 8: R16, 2007. DOI: 10.1186/gb-2007-8-2-r16 75

M. W. Smith, D.-F. Feng, R. F. Doolittle, Evolution by acquisition: The case for horizontal gene transfers, *Trends Biochem. Sci.*, 17: 489-493, 1992. DOI: 10.1016/0968-0004(92)90335-7 75

M. Poptsova and J. P. Gogarten JP, The power of phylogenetic approaches to detect horizontally transferred genes, *BMC Evolutionary Biology*, 45 p, 2007. DOI: 10.1186/1471-2148-7-45 76

R. Zhang and C. T. Zhang, A systematic method to identify genomic islands and its applications in analyzing the genomes of Corynebacterium glutamicum and Vibrio vulnificus CMCP6 chromosome I, *Bioinformatics*, 20(5):612-622, 2004. DOI: 10.1093/bioinformatics/btg453 76

S. Garcia-Vallve, J. Palau, A. Romeu, Horizontal gene transfer in glycosyl hydrolases inferred from codon usage in Escherichia coli and Bacillus subtilis, *Mol. Biol. Evol.*, 9: 1125-1134, 1999. 76

R. Merkl, SIGI: score-based identification of genomic islands, *BMC Bioinformatics*, 5:22, 2004. DOI: 10.1186/1471-2105-5-22 76

S. Karlin, Detecting anomalous gene clusters and pathogenicity islands in diverse bacterial genomes, *Trends in Microbiology*, 9: 335-343, 2001. DOI: 10.1016/S0966-842X(01)02079-0 76, 77

Q. Tu and D. Ding, Detecting pathogenicity islands and anomalous gene clusters by iterative discriminant analysis, *FEMS Microbiology Letters*, pp 269-275, 2003. DOI: 10.1016/S0378-1097(03)00204-0 76, 77

J. G. Lawrence and H. Ochman, Molecular archaeology of the Escherichia coli genome, *Proc Natl Acad Sci USA*, 95: 9413-9417, 1998. DOI: 10.1073/pnas.95.16.9413 76

W. S. Hayes and M. Borodovsky, How to interpret an anonymous bacterial genome: machine learning approach to gene identification, *Genome Res.*, 8: 1154-1171, 1998. DOI: 10.1101/gr.8.11.1154 76

W. Hsiao, I. Wan, S. J. Jones, F. S. Brinkman, IslandPath: aiding detection of genomic islands in prokaryotes, *Bioinformatics*, 19(3):418-420, 2003. DOI: 10.1093/bioinformatics/btg004 77

M. W. van Passel, A. Bart, H. H. Thygesen, A. C. Luyf, A. H. van Kampen, et al., An acquisition account of genomic islands based on genome signature comparisons, *BMC Genomics*, 6: 163, 2005. DOI: 10.1186/1471-2164-6-163 77

C. Dufraigne, B. Fertil, S. Lespinats, A. Giron, P. Deschavanne, Detection and characterization of horizontal transfers in prokaryotes using genomic signature, *Nucleic Acids Res,* 33: e6, 2005. DOI: 10.1093/nar/gni004 77, 80

N. J. Saunders, P. Boonmee, J. F. Peden, S. A. Jarvis, Inter-species horizontal transfer resulting in core-genome and niche-adaptive variation within Helicobacter pylori, *BMC Genomics,* 6:9, 2005. DOI: 10.1186/1471-2164-6-9 77

H. Ganesan, A. S. Rakitianskaia, C. F. Davenport, B. Tümmler, O. N. Reva, The SeqWord Genome Browser: an online tool for the identification and visualization of atypical regions of bacterial genomes through oligonucleotide usage, *BMC Bioinformatics,* 9:333, 2008. DOI: 10.1186/1471-2105-9-333 77

A. Tsirigos and I. Rigoutsos, A new computational method for the detection of horizontal gene transfer events, *Nucleic Acids Res,* 33: 922-933, 2005a. DOI: 10.1093/nar/gki187 77, 80

A. Tsirigos and I. Rigoutsos, A sensitive, support-vector-machine method for the detection of horizontal gene transfers in viral, archaeal and bacterial genomes, *Nucleic Acids Res,* 33:3699-3707, 2005b. DOI: 10.1093/nar/gki660 77

G. S. Vernikos and J. Parkhill, Interpolated variable order motifs for identification of horizontally acquired DNA: revisiting the Salmonella pathogenicity islands, *Bioinformatics,* 22, 2196-2203, 2006. DOI: 10.1093/bioinformatics/btl369 77

C. Dongsheng, C. Hockenbury, R. Marmelstein, K. Rasheed, Classification of genomic islands using decision trees and their ensemble algorithms, *BMC Genomics,* 11(Suppl 2): S1, 2010. DOI: 10.1186/1471-2164-11-S2-S1 77

G. S. Vernikos and J. Parkhill, Resolving the structural features of genomic islands: a machine learning approach, *Genome Res,* 18(2):331-42, 2008. DOI: 10.1101/gr.7004508 77

J. G. Lawrence and H. Ochman, Reconciling the many faces of lateral gene transfer, *Trends Microbiol,* 10:1-4, 2002. DOI: 10.1016/S0966-842X(01)02282-X 77

M. A. Ragan, T. J. Harlow, R. G. Beiko, Do different surrogate methods detect lateral genetic transfer events of different relative ages?, *Trends Microbiol,* 14:4-8, 2006. DOI: 10.1016/j.tim.2005.11.004 77

M. A. Ragan, On surrogate methods for detecting lateral gene transfer, *FEMS Microbiology letters,* 201: 187-191, 2001. DOI: 10.1111/j.1574-6968.2001.tb10755.x 78

J. Becq, C. Churlaud, P. Deschavanne, A benchmark of parametric methods for horizontal transfers detection, *PLoS ONE,* 5(4):e9989, 2010. DOI: 10.1371/journal.pone.0009989 78

112   BIBLIOGRAPHY

B. Linz, M. Schenker, P. Zhu, M. Achtman, Frequent interspecific genetic exchange between commensal Neisseriae and Neisseria meningitidis, *Mol Microbiol*, 36(5):1049-1058, 2000. DOI: 10.1046/j.1365-2958.2000.01932.x 79

E. P. C. Rocha, A. Danchin, A. Viari, Universal replication biases in bacteria, *Mol Microbiol*, 32: 11-16, 1999. DOI: 10.1046/j.1365-2958.1999.01334.x

B. Lafay, A. T. Lloyd, M. J. McLean, K. M. Devine, P. M. Sharp, K. H. Wolfe, Proteome composition and codon usage in spirochaetes: Species-specific and DNA strand-specific mutational biases, *Nucleic Acids Res.*, 27: 1642-1649, 1999. DOI: 10.1093/nar/27.7.1642 80

R. H. Baran and H. Ko, Detecting horizontally transferred and essential genes based on dinucleotide relative abundance, *DNA Res*, 15(5): 267-276, 2008. DOI: 10.1093/dnares/dsn021 80, 82

A. Muto A and S. Osawa, The guanine and cytosine content of genomic DNA and bacterial evolution, *Proc Natl Acad Sci USA*, 84:166-169, 1987. DOI: 10.1073/pnas.84.1.166 83

R. Sandberg, C.I. Bränden, I. Ernberg, and J. Cöster, Quantifying the species-specificity in genomic signatures codon choice, amino acid usage and G+C content, *Gene.* **42**, 311-335, (2003). 4, 5

E. Rivals, J.P. Delahaye, M. Dauchet, O. Delgrange, A guaranteed compression scheme for repetitive DNA sequences, *LIFL Lille I. University*, technical report IT **285**, 1995. 30

T. Coenye, D. Gevers, Y. Van de Peer, P. Vandamme, and J. Swings, Towards a prokaryotic genomic taxonomy, *FEMS Microbiol. Rev.* **29**(2):47–67, 2005.

Authors' Biography

OZKAN UFUK NALBANTOGLU

Ozkan Ufuk Nalbantoglu earned his B.S. in Electrical & Electronics Engineering from Bogazici University, Turkey, in 2004, and his Ph.D. in Engineering from the University of Nebraska-Lincoln, USA, in 2011, where he currently continues his research. His main research interest is in discovering the patterns in the organization of life and capturing trends in evolution. Specifically, he conducts computational studies of biological data with the concepts employed from information theory and signal processing. He is passionate about the inductive reasoning and corresponding engineering applications in life sciences, and he believes that this interdisciplinary paradigm shift will shape the face of science.

KHALID SAYOOD

Khalid Sayood received his BS and MS in Electrical Engineering from the University of Rochester, in 1977 and 1979, respectively, and his PhD in Electrical Engineering from Texas A&M University, in 1982. He joined the University of Nebraska in 1982 where he currently serves as the Heins Professor of Engineering. From 1995 to 1996, he served as the founding head of the Computer Vision and Image Processing group at the Turkish National Research Council Informatics Institute. His principal research interest is in how information is organized in data. He is the author of *Introduction to Data Compression* and the editor of the *Handbook of Lossless Compression*. He has also authored and co-authored a number of books published by Morgan-Claypool.

Printed in the United States
by Baker & Taylor Publisher Services